T0073900

INTERSTELLAR

ALSO BY AVI LOEB

Extraterrestrial:

The First Sign of Intelligent Life Beyond Earth

INTERSTELLAR

The Search for Extraterrestrial Life

and Our Future in the Stars

AVI LOEB

MARINER BOOKS

NEW YORK BOSTON

HarperCollins books may be purchased for educational, business, or sales promotional use. For information, please email the Special Markets Department at SPsales@harpercollins.com.

FIRST EDITION

Designed by Chloe Foster

Library of Congress Cataloging-in-Publication Data has been applied for.

ISBN 978-0-06-325087-1

23 24 25 26 27 LBC 5 4 3 2 1

To my three terrestrial partners, Ofrit, Klil, and Lotem,

and our neighbors in interstellar space

I am seeking to learn from a higher intelligence in outer space what we could aspire to be.

—Avi Loeb (February 26, 2023)

CONTENTS

INTRODUCTION

CONGRATULATIONS! YOU ARE AMONG the generations of humans to live at the dawn of humanity's interstellar future. We live in a time of great cosmological change, a time of mounting evidence that we are not alone, and that we can and will initiate encounters with the Universe, and whatever inhabits it, beyond our Solar system.

Over just the past decade the evidence of extraterrestrial life, extraterrestrial civilizations, and extraterrestrial interest in us has mounted rapidly. The possibility of life on Mars and Venus is being explored. The statistical likelihood of life existing on one of the innumerable exoplanets in a star's habitable zone is high, and soon to be explored by spacecraft capable of sending data back within a human lifetime. Most important, for the first time, the search for near-Earth extraterrestrial artifacts is the work of science, privately and publicly funded. Whether or not humanity persists long enough to get off its home planet and to exist independent of its home star is on us. And, if we are diligent, smart, and intrepid, just maybe we manage this with an extraterrestrial assist.

One opportunity to do so has already slipped through humanity's collective fingers.

Data support the possibility that in 2017 an extraterrestrial-manufactured artifact passed through the Solar system. That year, astronomers, using data collected by the Panoramic Survey Telescope and Rapid Response System (Pan-STARRS) located at

Haleakala Observatory, Hawaii, identified an interstellar object that they called ‘Oumuamua, which translates to “scout” in Hawaiian. Based on the wealth of empirical data collected about the object, I argued that it was most plausibly of extraterrestrial manufacture, rather than a naturally occurring interstellar rock.

The data revealed that ‘Oumuamua’s shape—long and remarkably flat—was to an extreme degree unlike any space object seen before. The data also showed to a certainty that its unusual trajectory around the Sun was changed, not by any visible outgassing such as occurs with all comets, but most likely due to it being pushed by solar radiation, just like NASA’s rocket booster, 2020 SO, which was discovered by Pan-STARRS on September 17, 2020. Then there was ‘Oumuamua’s extraordinarily low velocity at the point when our Solar system encountered it, which was measured at what astronomers call the Local Standard of Rest (LSR). In space everything is moving relative to everything else. An object at “rest,” such as ‘Oumuamua, is an object with a velocity that makes it comparatively still among all that movement. This is rare. ‘Oumuamua’s being at the LSR made it an outlier to 99.8% of all stars. Nature infrequently puts objects at the LSR. If, however, humans wished to manufacture an object and place it at the LSR, that would be well within our technological know-how. And that is why I likened ‘Oumuamua to a buoy our Solar system ran into rather than a rocket aimed at our Solar system.

Because we had no means of imaging ‘Oumuamua, because we had no means of capturing ‘Oumuamua, and because our instruments of detection were not designed and built for detecting near-Earth objects of potential extraterrestrial manufacture, we have only inconclusive data. Inconclusive data are, of course, the case for much of science, and it is the presumptive condition of all science at the leading edge of discovery in new domains of inquiry. It is also the case, as I argued in *Extraterrestrial: The First Sign of Intelligent Life Beyond Earth*, that of all the explanations for the data we did manage to collect concerning ‘Oumuamua, the simplest, most plausible explanation for its behavior was that it was manufactured, albeit not by humans.

Simple and plausible hold, of course, only if you grant human civilization is likely not alone in the Universe. For many, granting that possibility has proven not a stumbling block so much as a brick wall. To date, however, scientists seeking a natural explanation for 'Oumuamua's extreme elongation and smooth non-gravitational acceleration have encountered only more walls. Briefly consider Jennifer Bergner's and Darry Seligman's "Acceleration of 1I/'Oumuamua from radiolytically produced H_2 in H_2O ice," published by *Nature* in March 2023. It offers a refinement of Seligman's 2020 hypothesis that 'Oumuamua was composed entirely of molecular hydrogen ice, which, along with the hypothesis that 'Oumuamua was pure water or pure nitrogen, was subsequently acknowledged to be unworkable. However, cosmic ice made of water ice partly dissociated into hydrogen by cosmic rays in interstellar space, Bergner and Seligman proposed in 2023, might work. Perhaps, but neither simply nor plausibly. First, we have no data that suggests nature produces interstellar icebergs made up of water and hydrogen. And second, the evaporation of hydrogen trapped in Bergner's and Seligman's proposed water-hydrogen comet wouldn't explain 'Oumuamua's observed non-gravitational acceleration. The models Bergner and Seligman advanced ignore the cooling that results from evaporating hydrogen, which once accounted for decreases the hypothesized outgassing by a factor of 3, which makes the model untenable as an explanation for 'Oumuamua's known properties.

That five years after 'Oumuamua's discovery the scientific debate about its possible origins remains robustly unsettled is good news for science. That an extraterrestrial origin is for many still a nonstarter is sobering news for human civilization.

Since the publication of *Extraterrestrial* I have been asked many times about "Day Two," or what I believe will happen immediately after humanity encounters evidence of extraterrestrial technological civilizations. We already have the answer. The media will take notice. A small percentage of the public will take notice. But an overwhelming majority of humans will continue to live their lives without giving it much thought. Our politicians will continue to seek re-election, those in business will continue to seek profits, and Day Two will look and feel much as Day Minus One.

This would not surprise me. What was a little more surprising—and disappointing—was the fact that on Day Two, and continuing over Years One through Five, a majority of the scientific community expressed a skepticism about this evidence that was greater than anything directed at scientific speculations such as string theory, types of dark matter, and multiverses. This is despite the fact that to date we have no empirical data demonstrating a theorized string, a dark matter particle, or a single universe other than our own. Scientists, in other words, are more comfortable asserting the existence of phenomena they have no empirical evidence for than accepting the possible existence of a phenomenon—Extraterrestrial Civilization (ETC)—for which we do.

What this all rolls up to is the following realization: The question to answer isn't, "What will we do on the day after we encounter evidence of ETC?" The question is, "What will we do as we encounter ever more evidence of ETC?" And that is why I have written *Interstellar.*

Humanity is on the cusp of profound discoveries about our cosmic neighborhood. The Universe is knocking on our door just as we are preparing to open it, and the great likelihood is that incontrovertible evidence of extraterrestrial sentient intelligence is just on the threshold. Cosmological-firsts and civilization-altering realizations are perhaps even just months away.

We need to prepare. We need new instruments of observation and interception. We need far greater transparency, and with it coordination of effort, among scientists and governments. Arising out of that transparency, humanity needs new expectations about how encounters with interstellar objects, and eventually with non-terrestrial life and non-terrestrial civilizations, are likely to unfold. How we prepare, how we as a civilization and a scientific community now behave, will determine the psychological costs and benefits of this new era of sentient terrestrial existence. For starters, we will need a new vocabulary.

The rapid pace at which evidence of the plausible existence of extraterrestrials can accumulate is captured by the fact that in 2022 the United States government admitted the existence of unidentified aerial phenomena, or UAP, previously known as UFOs. Notably, almost immediately after admitting UAP existed, America's military began firing missiles at them. Within the first two months of 2023, American planes destroyed a high-altitude balloon, of known Chinese manufacture, and three unidentified objects. That they were most likely human-made doesn't change the fact that we fired on them without being certain. Just months earlier, and for the first time in our history, humanity began scientifically rigorous multipronged initiatives to search for alien artifacts. The first of these, the Galileo Project, which I direct, launched slightly before the government's acknowledgment of UAP, and by 2022 the first of the Project's observatories specifically designed to identify UAP went live. And within 2022 alone, 'Oumuamua went from being the only identified interstellar object to being chronologically the third of four.

Within that year, a review undertaken by my student Amir Siraj and I of the catalog of fireball and bolide data of the Center for Near Earth Object Studies (CNEOS) was determined to have identified two interstellar meteors. Both of them, the data show, are built of far tougher materials than the vast majority of space rocks that originate in our Solar system. And one of those two awaits our recovery, resting at the bottom of the Pacific Ocean just north of Manus Island. Even without full access to all the data the United States government holds, we know to within a ten-kilometer grid where its fragments rest. During 2023, an expedition I am heading will, it is hoped, discover fragments of an interstellar object.

No matter what is discovered on that ocean floor, as a civilization we need to better prepare ourselves.

Whether we acknowledge it or not, my generation and yours, as well as the ones that will follow, are living in the first years of a new era, one in which humanity steps into its interstellar future.

We have never been so close to scientifically valid proof that life on Earth and human civilization are not alone in the Universe. I am convinced that we are tantalizingly close not only to learning terrestrial life is not the only life in the Solar system, and that human civilization is not the only civilization to exist or have existed: I am also convinced that most of humanity is not ready.

Decades of science fiction has ill served our civilization. In virtually all of these stories, space and extraterrestrial life are used as backdrops in front of which are performed familiar genre narratives, be they horror, fantasies, love stories, or action-fests. Among the possible outcomes of our first contact with ETC, the least likely is a human-alien handshake in front of the White House or a contest of our missiles versus their lasers. For reasons of science and statistics, the far greater likelihood is that we will encounter ETC garbage or an extraterrestrial AI probe, and the encounter will occur at all only because we seek it out.

Along with building new tools and defining, and thereafter funding, new scientific and technological research, humanity needs to reset our cultural understanding and expectation of what it *means* to encounter an extraterrestrial object. And at the leading edge of research and theory, astrophysicists, cosmologists, and astronomers are gathering data-informed insights into humanity's meaning and purpose in the 13.8-billion-year-old Universe.

Embracing the fact that UAP exist, that interstellar material of exceedingly rare material strength is within our grasp, and that evidence of ETCs is going to rapidly mount is to confront the near and far future possibilities of humanity. The times we live in demand that we undertake this search with all the intellect, skill, dedication, and joy we can muster. Not just because we now have evidence that UAP are real, and that some percent of them are perceived threats, and not just because ETCs are increasingly irrefutable, but because we live at a time of ever-mounting alarm over human-made existential threats, from climate change to war. All humans must take up this work because humanity stands at a crossroads. The next steps we, and especially our scientists, take, I am convinced, will decide if our civilization is destined to persist, spreading terrestrial life throughout the Universe, and thereafter where humanity stands in the pantheon of cosmic civilizations.

It is tempting, and humanity's hubristic default, to presume we warrant a place within that pantheon. The greater statistical probability is that we don't.

The best-known scale for assessing civilizations was invented in 1964 by the Soviet astronomer Nikolai Kardashev. Kardashev dedicated much of his career to the theory and practice of searching for extraterrestrial life, but arguably his best-known contribution is the Kardashev Scale. By this scale a civilization that has mastery of its planet's energy resources is labeled a Type 1. That is about where humanity currently stands. Once a civilization has mastered how to harness the total energy output of its host star it is elevated to a Type 2. And a Type 3 civilization has learned to draw on the full energy of its galaxy.

The Kardashev Scale has merits, but it has, I believe, one clear deficit. Energy consumption tells us too little about the culture of a civilization, its priorities, its ambitions, its expectations for itself and life on and off its home planet. I believe there is a more useful perspective. The technological level of civilizations should not be gauged by how much power they tap. Instead, it should be measured by the ability of a civilization to reproduce the astrophysical conditions that led to its existence.

By this cosmic scale, a *C-class* civilization is one that is able to recreate the habitable conditions on its planet without relying on the energy of its home star. For example, had the dinosaurs attained this level of technological ability, then the consequences of the Chicxulub impactor, an asteroid or comet ten miles wide that struck Earth 66 million years ago, would have been devastating but not an extermination-level event. While in 2022 NASA's Double Asteroid Redirection Test (DART) was the first demonstration of planetary defense technology, or the ability to impact and thereby possibly divert an Earth-bound asteroid or comet, our current ability to nudge anything the size of Chicxulub is zero. Nevertheless, we can imagine some fraction of humanity, facing a similar crisis, surviving off of nuclear energy and food grown in greenhouses.

A *B-class* civilization could adjust the habitable conditions in its immediate environment to be independent of its host planet and host star. A human civilization at a *B-class* level would be capable of building a Noah's ark in fact. Spreading first throughout the Solar system, then beyond it, and most likely relying upon AI probes with the programming and technology capable of restarting terrestrial life on distant exoplanets, human-built spacearks would hedge against the destruction of Earth and Sun. And at the top rung of the ladder would be an *A-class* civilization. This is one capable of recreating the cosmic conditions that gave rise to its existence, namely a civilization capable of producing a baby universe in a laboratory.

Achieving the distinction of an *A-class* civilization is plausible by the measures of physics as-we-know-it. The related challenges were already discussed in 1990 in a paper co-authored by Edward Farhi, Alan Guth, and Jemal Guven. Their assertion: "We are suggesting . . . that the laws of physics as we know them permit in principle the creation of a new universe by human initiative." The language of math applied to the known physical laws points us toward the testable hypothesis that the opening phrase of the Book of Genesis might more accurately be written, "In the beginning . . . there was the lab coat."

This leads me to a more difficult question: Where, as of 2023, would humanity fit on this cosmic scale of civilizations? I believe we're closer to a *D-class* civilization, or one actively degrading its home planet's ability to sustain conditions that prolong life and civilization.

And now we can begin to see our predicament. We also begin to see our opportunities. I am convinced that if we are to seize them all humanity must learn to lean into science. It is science that will give us a new vocabulary for the upward trajectory of our civilization, science that will guide technology to give us the necessary tools, and science that will best determine why the psychological costs of encounters with ETC will become, in practice, benefits.

A goal of this book is to make and keep you excited about our interstellar future. I believe humanity's ability to sustain an optimistic and scientific focus on the possibilities inherent in the reality shared by all sentient intelligence will determine whether or not our civilization joins the ranks of interstellar civilizations.

To accomplish its goal, the book is split into two parts. The first five chapters speak more directly to the steps taken and still ahead of us to fully embrace what I think of as humanity's future after yesterday. In these pages, I will address the practical measures our civilization must take along multiple fronts to expand our search for extraterrestrial artifacts, to prepare our civilization for their discovery, to consider the technology we have and will soon have to advance the science behind our search. These steps are either informed by science or dependent on science. What we must do the moment after we discover evidence that we are not alone has already become the work we are and will need to continue doing.

Our interstellar present raises practical questions of science and technology. New telescopes will be built. New spacecraft will be designed and launched. These will speed the discovery of nonterrestrial life, artifacts humans did not manufacture, evidence of other civilizations. The timing of these discoveries is unknowable. Certain, however, is that humanity becoming interstellar raises not just practical questions, but touches on the most fundamental questions of our sentient existence. Already, the practical questions cannot be separated from these more philosophical questions. That is why the last five chapters of *Interstellar* are more wide-ranging, even somewhat spiritual. I believe our interstellar future scientifically raises, and helps answer, the most profound questions conscious intelligence confronts. It is in the spirit of an optimistic, scientific, philosophical enthusiasm that I hope before 2023 is over to find and hold evidence of an extraterrestrial artifact.

And maybe, just maybe, when I do find it, it will have a button or two that I can push.

PART I

1

ASCENDING THE LADDER OF CIVILIZATIONS

IN JUNE 2021, THE US Department of Defense delivered a report to Congress confirming that military pilots have seen and photographed UAP, that these objects are perfectly real, and that despite decades of gathering data on the phenomenon, the Department of Defense does not know what the objects are. Promptly called "The Pentagon Report" when it was released, it remains one of the more significant indicators that humanity now lives during the early years of a new era.

February 4, 2023, established that our transition to this new era will be bumpy. On that date, an aerial phenomenon we could explain, a Chinese-manufactured balloon, was shot down by US Air Force pilots off the coast of South Carolina. China declared it had been built for meteorological research purposes. The United States made an ever more evidence-backed claim that it had been built for espionage. Within weeks, three UAP were shot down. Because of where they were shot down—over Alaska, Canada's Yukon Territory, and over Michigan's Lake Huron—gathering evidence as to their manufacture has been difficult. But the scientific presumption is that all three UAP were human-made.

We do not know if the United States military's admission of

the existence of UAP was an early acknowledgment, perhaps even a signal to its adversaries, that it knew some percent of reported UAP were surveillance technology launched by other nations here on Earth. Evident in subsequent news conferences, given by both the military and the White House, was confusion. In one instance, it took two missiles to destroy a UAP, the first having missed the small object. And at a press briefing on February 13, 2023, White House Press Secretary Karine Jean-Pierre declared, "I know there have been questions and concerns about this, but there is no—again no—indication of aliens or extraterrestrial activity with these recent takedowns," all but begging the unanswered question, What would such an indication be? Tipping a hand to its own uncertainty, the Department of Defense signaled it was broadening its scope of interest when it renamed the "Airborne Object Identification and Management Group" the "All-domain Anomaly Resolution Office." From curiosity with just airborne UAP, the government is indicating by the name change that this new office's concern will be with unexplained phenomena being tracked everywhere—space, air, land, and sea. In less than two years, the American government went from denying, even belittling, interest in UFOs to rebranding them UAP, declaring them a pressing matter of national concern, and on occasion ordering the military to neutralize some of them.

How quickly worries about UAP shift from being the focus of one nation's military to a species-wide scientific concern will be one measure of how quickly humanity might ascend the ladder of civilizations.

The confusions of February 2023 were hinted at by those of October 2022. I was then reminded that this shift of perspective, from narrow nationalist interests to a shared interstellar curiosity, will not happen seamlessly. That was when I first read that Ukrainian astronomers were reporting multiple sightings of "dark" objects three to twelve meters in size, reaching speeds of fifteen kilometers per second, and traveling with no optical emissions. I was at home, sitting at my desk, when a string of emails called my attention to the reports by way of asking me for my thoughts and reactions. Most of these emails led by pointing out that, if what was being reported was true, the capabilities of these "phantom" objects exceeded anything human-made aircraft or rockets were capable of. Were Ukrainians seeing UAP that evidenced an extraterrestrial civilization's interest in terrestrial events?

Most of my replies led by pointing out what we knew for certain. Ukraine was in a military conflict with Russia, and this guaranteed the introduction of a lot of noise—ordnance, drones, airplanes—in the astronomers' data. In short, the lack of reliable data made the reported phantoms an unpromising phenomenon for science to address. And so I ignored the reported sightings until a high-level official in the United States government showed up at my front door and asked for my thoughts on them. A few hours later, I wrote a paper laying out the scientific reasons the Ukrainian reports were most likely false. If we accept the reported distance and speed of the dark objects, physics and math tell us that the friction of any dark object that blocks light with the surrounding air would generate a bright optical fireball. If, however, you reduce the objects' inferred distance by a factor of ten, then they would be fully consistent with the size, speed, and appearance of artillery shells.

On reading my article, the Ukrainian astronomers who authored the original report declared me a "theorist" and stuck to their belief that the objects remained mysterious UAP, and, striking a curious note of national pride, they observed, "our characteristics of the objects are very similar to those of US military pilots and Canadian civilian pilots."

Hidden in the brief global interest with phantom UAP possibly in the skies over war-torn Ukraine and in the United States destroying UAP of uncertain origin is, I think, a larger worry. Humanity runs the risk of spending years, perhaps generations, couching its interest in extraterrestrial artifacts within historically and locally defined contests among terrestrial nations. It is in this sense that we need a new vocabulary that reflects an interstellar frame for understanding our species' predicament. For example, hidden in the exchanges that followed the Ukrainian report was an unasked question: Why do humans imagine that so advanced and technologically capable extraterrestrial civilizations would *want* to send phantom craft to our atmosphere to observe our rather parochial practice of industrial murder? Isn't it more probable that advanced ETC would be animated by the broader scientific curiosity the Chinese want to claim was the purpose of their balloon?

We must similarly ask, why is it that the United States' concern with UAP has always been, and largely still remains, the province of branches of the American military? The Pentagon Report was momentous. A wealth of empirical data about UAP was finally acknowledged to exist; more, still classified data very likely exist. That after the first balloon was spotted over the United States, the North American Aerospace Defense Command (NORAD) stated it had enhanced its radar to pick up on smaller objects suggests they had been able to do so all along. What the nations of the world do with the data they hold, and to which branches of government they entrust those data, goes immediately to the question, what will humanity do with the mounting empirical evidence that UAP exist? How we respond, and with what guiding assumptions, will, I believe, determine for how long we remain a *D-class* civilization being led by our local history rather than by the methods and logics of universally shared scientific inquiry.

We did not need the Pentagon Report, Ukrainian astronomers, and NORAD to tell us that UAP are not science fiction. They are simply science. Explaining the physical world's phenomena is what the scientific method was invented for. The Universe, of course, is full of unexplained and insufficiently explained phenomena. Some remain squarely in the realm of theory—dark matter, for just one example. Others, like UAP, are questions answerable through the collection of more and better observational data. What is more, we know manufactured technology arriving from interstellar space is plausible because we've already sent human-manufactured technology into interstellar space. The effort to answer whether another civilization has done likewise, and perhaps with far greater sophistication, quickly presents us with some of the most awesome, enduring questions of human civilization.

The conclusive, empirical discovery of a vast extent of "dark" matter previously invisible to us would be significant. It is unlikely, however, to alter our understanding of ourselves. By contrast, whatever we discover through the science of extraterrestrial artifacts—say, that life and civilizations are common throughout the Universe or life and civilizations are exceedingly rare—we will nevertheless confront a species- and civilization-altering question: What is humanity's place among life in the universe? There is something frustrating about the fact that for decades humanity has expended far greater amounts of money and talent in pursuit of a better understanding of, and some empirical evidence for, dark matter than it has in pursuit of extraterrestrial artifacts. But there is a silver lining. There is much to be done that will accumulate evidence touching on the existence of ETC capable of sending UAP to Earth's atmosphere, and some of it can be done quickly.

Whether or not this scientific work is done quickly depends on how long it takes us to redress how ill-prepared we are to undertake the work of resolving UAP into EAP (explained aerial phenomena). And that will require, I believe, confronting an uncomfortable cultural fact: like our planet, our *D-class* civilization is immature.

In the family history of all the stars in the Universe, our own is a toddler. Most stars were born billions of years before the Sun. Many are so old they have already consumed their nuclear fuel and cooled off to a compact Earth-sized remnant known as a white dwarf. And of the billions of stars in our own Milky Way galaxy, roughly half host an Earth-sized planet in the zone that would allow for liquid water and the chemistry of life.

That means that here in the Milky Way alone, the dice of life have been rolled in innumerable places under conditions similar to those that produced life on Earth, and likely much, much earlier.

Surely, then, it is reasonable to suspect that life elsewhere not only exists but is quite common, and that in many cases it began long, long before our microbial ancestors got their start here on Earth. If so, then intelligent species may well be billions of years ahead of us in the building of technological civilizations. In addition, during the very short time span that encompasses human history, entire civilizations—Mesopotamian, Roman, Aztec—have come and gone and now exist only in cultural echoes that merely a few scholars hear, and through archaeological digs. Scale this analogy to the Universe and we can postulate the statistically plausible bet that other civilizations have come and gone, some leaving discoverable echoes and artifacts. And that the few civilizations that might persist could be millions, perhaps billions, of years older than our own.

A THOUGHT EXPERIMENT

Because we have already launched our own interstellar craft, we can usefully turn the tables on our own search for evidence of another interstellar civilization. At the time I type this, Voyagers 1 and 2 are the only spacecraft composing NASA's interstellar mission; NASA engineered, built, and launched craft in expectation of their discovery by another sentient intelligence. Imagine, a million years from now in a galaxy far, far away, that NASA's Voyager 1 space probe is diverted by the gravitational pull of a star, where it draws the attention of a species on one of the Universe's innumerable planets that exist in a star's habitable zone.

Voyager 1 is about to fulfill NASA's fondest dreams for it, and remarkably, Voyager 1 has, despite its long interstellar journey, survived largely undamaged since its launch in 1977. Its decahedral bus, a mere 47 centimeters in height and 1.78 meters across, has miraculously weathered its million-year journey intact and its most visible feature, a 3.66-meter-diameter parabolic antenna, is unharmed. Voyager 1 is pulled into a solar system's orbit on a trajectory that has it pass close enough to the species' telescopes that it is detected by the starlight it reflects. And in short order, the data streaming in from their other instruments help them determine

Voyager 1's speed, rotation, rough point of origin, heat signature, and clues as to its material composition.

Imagining the means by which Voyager 1 might be identified by another technologically adept civilization is less speculative than you might think. Whatever the biology of the species, their observatories would not look much different than ours. The Universe's fixed physical laws effectively serve to put boundaries on the design of devices used to study them and tease out their consequences. There are only a finite number of ways you can build cameras that photograph distant objects and telescopes that pick up infrared wavelengths, just as there are a finite number of ways to generate the electricity necessary to power them. Put differently, all civilizations that confront the same unchanging natural laws and attempt to understand those laws will be constrained by those same laws as they manufacture technology, and they will tend toward certain common understandings, design choices, and conclusions.

True, the gap in technological capabilities between civilizations could be so great that one might look upon the technology of the other with as much comprehension as a chimpanzee watching a presentation at an Apple launch event. Humanity confronted with a civilization that had the benefit of millions of years' experience of total mastery of all the physical laws, classical and quantum, would be akin to such a chimpanzee, if not something even less flattering, say a termite. But it is also possible that the gap between species could be small enough to make the relationship more that of a student to a teacher.

Being generous, we might posit such a gap to be about 2,000 human-history years long. I imagine that the inventor of the Antikythera Mechanism, a shoebox-sized device about two millennia old that ancient Greeks used to calculate the locations of celestial bodies, on being introduced to an iPhone might still think of the forces of the natural world in terms of gods and spirits, but would also bring a remotely usable frame of reference to the table—far more than a chimpanzee or a termite. Even a student so disadvantaged could make rapid progress understanding an iPhone, especially with a sufficiently gifted and patient teacher.

Enough common knowledge and even a two-thousand-year gap is likely bridgeable. Of course, if a species is without sufficient scientific knowledge, or is resolutely interested only in its own planetary biology and civilizations, or elects to be incurious about the possibility of other planetary civilizations, then it wouldn't have a clue that a new piece of extraterrestrial trash had intruded on the orbit of their sun. Voyager 1 would come and go and they would continue on with their governing myths and, very likely, the presumption of being both the smartest and the only civilized life in the vast expanse of the Universe.

For the sake of the current thought experiment, we must assume that this hypothetical species in the distant future is scientifically capable and interested in the arrival of even something as small and unassuming as Voyager 1 passing 24 million kilometers, or about the distance Venus is from Earth, from their home planet.

Here is my first question: How impressed should we be with this species' scientific and technological prowess?

The answer: enviously impressed, but not for the reason you'd think. Humanity currently doesn't have equipment capable of detecting a probe of that size at that distance. 'Oumuamua was at a similar distance from Earth, but it had a width and length of hundreds of meters (being most probably pancake shaped, its depth was harder to determine and can be left as "thin"). That's considerably larger than Voyager 1, roughly a football field compared to a diminutive football player. And while the astronomers operating Pan-STARRS telescopes atop Hawaii's Haleakalā Observatory in 2017 were able to discover this interstellar object, those aiming the infrared Spitzer Space Telescope—designed to peer deep into space rather than at rapidly passing nearer objects—failed to detect any heat signature. Whether there was none, or the signature was simply below the telescope's ability to detect, we do not know. No instrument in humanity's arsenal of cosmic discovery was then built for the purpose of detecting anomalous, and perhaps extraterrestrial, interstellar objects.

The species tracking Voyager 1 is more fortunate, possessing instruments capable of alerting them to the probe's existence. And if this species has telescopes sophisticated enough to spy Voyager 1 in transit, that suggests strongly they were looking for just such an object. And in turn, that means the odds are good that their civilization also possesses the means to intercept, image, and even capture Voyager 1. And if all of these presumptions hold, then the species would soon be the possessors of the twelve-inch gold-plated copper disk encased in a protective jacket and mounted on the side of Voyager's bus. Soon, the species' scientists, and perhaps its planet-wide population, would hear a soundtrack that includes words of welcome in fifty-five terrestrial languages, diverse recorded sounds (thunder, a human heartbeat, crickets, the liftoff of the Saturn V rocket, a mother kissing her child), and an eclectic collection of music. More important, they would possess incontrovertible evidence that they are not alone in the Universe.

Unfortunately, to date humanity does not have telescopes capable of picking up the existence of an interstellar object the size and distance posited for Voyager 1. And among our rapidly increasing fleet of spacecraft, we still have *nothing* capable of chasing, intercepting, and photographing, let alone capturing, any such object.

This unfortunate fact leads to my second question: How great is the scientific gap between this species' civilization and humanity's? In fact, the technological gap is minimal to nonexistent. We possess the technology to create such instruments. The reason humanity does not currently have them is not a matter of means, but a matter of choice. And this is the real reason we should envy this hypothetical species.

Yes, we would have much to learn from a species capable of bringing Voyager back for show-and-tell in their planetary museums. But the nature of those lessons wouldn't be primarily technological or scientific. They would have more to do with the relative difference between the species' place on the cosmic ladder of civilizations versus humanity's.

Consider that for decades we've known that manufacturing and launching interstellar craft is not difficult. After all, we did it in 1977, sending both Voyagers into space at the reasonable cost of $865 million. (That was also the year humans in the millions packed theaters to watch *Star Wars*, which happened to gross around $775 million.) If we can do this, another planetary civilization could, and depending on their priorities they might have created an interstellar mission able to send out far more than just two craft. And not just priorities, but longevity: The first words that open *Star Wars* are the only ones that make any of its make-believe even remotely plausible. "A long time ago" trailed across the screen before "In a galaxy far, far away. . . ." This was director George Lucas's subtle reminder to his viewers that the Sun formed only 4.6 billion years ago, in the *last third* of cosmic history. If human civilization manages to persist, if the dawn of our interstellar era is not also its dusk, we will manufacture ships that can chase, image, and even capture a UAP or interstellar object.

The reasons we are unable to do so now is that as a civilization we have not prioritized the search for extraterrestrial technology. More is at stake, however, than missing the momentous discovery that we are not alone. We are currently a *D-class* civilization, characterized by our short-sighted pursuit of selfish goals at the expense of the long-term survival of our civilization and our planet. If humanity fails to ascend the cosmic scale of civilizations, the odds increase that we will perish even before the Sun does. To jolt us out of our current complacency, humanity needs a sufficiently grand ambition. That ambition, I am convinced, is the scientific and technological pursuit of proof of an ETC.

A NEED FOR NEW LEADERSHIP

Scientists must and will be at the forefront of positioning humanity to discover life and civilizations in the Universe. And that is why we need to think first of politicians. Before we, and *Interstellar*, can turn to the science, it is essential that we confront the stumbling blocks to science's rapid advance. Well before any alien shows up to say, "Take me to your leader," humans must ask, "Which of our leaders are preparing us for the future to come?"

We must choose politicians with the curiosity and imagination to prepare our civilization for this endeavor. This is for a practical reason. However important the space initiatives of wealthy private citizens, it's the politicians who will make the decisions and award the significant funds to mount a concerted search for evidence of extraterrestrial civilizations. It is also for a more subtle reason. It will be the leaders of Earth's nations who will adopt the attitudes and advance the policies to prepare humanity for its ascent up the ladder of civilizations. Fortunately, that work has already commenced. Even if it was just recently true, and in many parts of our planet still true, that advocates for seeking extraterrestrial artifacts were relegated to the aluminum-foil-hat wearing, presumptive crazies, some leaders have already been at work destigmatizing the search for extraterrestrial civilizations and technology. That the Pentagon Report was released at all is in part due to their efforts. This is why, as humans across the globe exercise what authority they are allowed to exercise in choosing their leaders, I recommend we bear in mind the example of the late Senator Harry Reid of Nevada.

It was Senator Reid—with Senators Ted Stevens, a Republican of Alaska, and Daniel Inouye, a Democrat of Hawaii—who in 2007 secured the relatively meager sum of $22 million to pay for the clandestine Advanced Aerospace Threat Identification Program. That was an important step toward the Pentagon's report on UAP in 2021. But it was Reid's frame of mind, even more than his efforts to recast a secretive, stigmatizing dialogue into a public, scientific one, that should inspire us. He was curious. He was open. He was courageous. When, in 1996, a friend suggested he attend a serious discussion of UFOs, Reid went and was fascinated by what he heard. When his friend John Glenn, the Mercury astronaut elected to the Senate, told Reid that he, too, was curious about the possibilities of life elsewhere, Reid was impressed. When his staff told him, "Stay the hell away from this," Reid ignored them.

Senator Reid had no Ph.D. Brought up in a house made of railroad ties in the tiny Nevada town of Searchlight, he hitchhiked forty miles to high school, where the science classes were few. But at night he would lie in the yard on an old mattress and watch the shooting stars, wondering how the Sun stayed so hot. He had a scientist's curiosity. It never left him. When he had the power to help with the search for evidence of life elsewhere, he used it. "I have never intended to prove that life beyond Earth exists," he wrote not long after he left the Senate. "But if science proves it does, I have no problem with that. Because the more I learn, the more I realize that there's still so much I don't know."

The vocabulary humanity needs now will follow the frame of exploration and explanation our leaders provide. Curiosity, a willingness to be guided by a hypothesis and not a wished-for outcome, plus sufficient trust to invite the public into the pursuit of proofs significant not just to a nation or political party but to all humans promise a path forward for our civilization. The more hidebound our planet's leaders and governments, the more uncertain that path.

I was reminded of this in early 2022 during a forum with former Secretary of State Henry Kissinger. The ninety-eight-year-old politician and diplomat brought the full weight, and all its associations, of the twentieth century to the conversation. I asked him the following question: "If we find extraterrestrials, how do we play 'realpolitik' with them when we know nothing about their society?" Realpolitik, or the approach to the gamesmanship among terrestrial nations, was predicated on understanding in a pragmatic way the goals and interests of your nation's counterparts, ally or adversary. How, I was asking, can that approach work when we confront a civilization about which we know absolutely nothing? Kissinger is considered one of the grandmasters of realpolitik—and his response to my question was telling:

> This is a very good question. If we ever establish contact with an extraterrestrial society, we should make huge efforts to get into some dialogue to understand their perception of what they are facing and what needs to be done, and then see to what degree it is compatible with our perception, including artificial intelligence and cyber technology. Our civilization might not survive if a war is conducted to the limits of each side. We should prepare to analyze issues as they arise and understand, as is the case with Russia or China, how do other societies view reality, what can we learn, and what do we want to achieve after understanding them in a general context.

There is wisdom here, but of a distinctly terrestrial sort. Secretary Kissinger's use of Russia and China as the case studies to consider, his echo of the Cold War worry of an unsurvivable war, and his faith that the general context of shared understanding will be how all societies, terrestrial and not, view reality are all telling. They reflect consideration of the metaphorical gravity of human civilization rather than the physical laws of gravity itself. Those universal laws are far more likely to be our common vocabulary for communication.

If we are to survive, let alone benefit from, first contact with an extraterrestrial technological civilization, we will need leaders who think like Reid not only in the United States but in every country. The search for extraterrestrial civilizations cannot be mounted by a nation here and a nation there. The discovery of extraterrestrial artifacts cannot become the classified property of our historically bound, narrowly interested twenty-first-century nation-states. The search for, and any discovery of, ETC needs to reflect universal principles and methods. As Kissinger makes clear, this is neither obvious nor easy. The most expedited progress, however, will occur if how we seek the data and how we manage any encounters follow the protocols of the scientific method, of the transparency of peer review, with "peer" understood as the world's competent scientific minds.

To the present, most terrestrial governments have confronted the question of encountering extraterrestrials as Secretary Kissinger did. They scale it as a terrestrial problem, a familiar concern, one that participates in the contests of states and ideologies that have governed so much of the last centuries of human existence. This is the wrong frame. We must consider it as parallel to responding to a global pandemic. Just as in the struggle to combat a pandemic, we must think and act as a species, not as a hodgepodge of individual nations. And just as with the recent COVID-19 pandemic, our most hopeful way forward is to support, trust, and encourage science.

CIVILIZATION AT A CROSSROADS

When the Pentagon released its report of evidence of UAP, listing the many unanswered questions and citing the absence of high-quality data, I expected my fellow scientists to respond as we are trained to do from the moment we first step into a lab: to begin the work of transparently gathering more, and more useful, data. After all, that is squarely in our professional wheelhouse. The raison d'être of scientific work is the exercise of the scientific method—gather data, hypothesize an explanation, test your hypothesis, report results and conclusions, gather more data, and repeat. To any scientist, I thought, the Pentagon Report would be heard as a loud cry for more data.

I was dismayed to see many scientists, including astrophysicists and devoted advocates of the Search for Extraterrestrial Intelligence (SETI), respond with a collective shrug. Many said or inferred that UAP were very likely traceable to common terrestrial phenomena—drones and weather balloons—and not worth time and effort. Many viewed the subsequent destruction of UAP by the US Air Force as confirmation of their presumptions. These sightings documented by the Pentagon and the takedowns by the military, they said, did not constitute the "extraordinary evidence" that the astronomer Carl Sagan famously demanded before he would be willing to countenance the possibility of extraterrestrial life. And that was that.

Like Galileo's critics, they chose not to look through new telescopes constructed for the purpose of collecting evidence on this unexplained phenomenon.

The Pentagon knew better. Deep in the text of its momentous report, without even touching upon 'Oumuamua, the writers acknowledged that the larger problems and possibilities presented by UAP now belong on the desks of scientists. However well-trained in their own field, even the most sophisticated aircraft pilots are not scientists. Their testimony, while unquestionably valuable, is vulnerable to the emotion and subjectivity inherent in any human experience, let alone the high-anxiety moment of seeing an object in the sky that fits no known description. Their equipment, however advanced for military purposes, is designed for tasks other than recording data of UAP. Such observers also face the fear of scorn and ridicule from their colleagues and superiors. When *60 Minutes* interviewed former Navy pilots Commander Dave Fravor and Lieutenant Commander Alex Dietrich about their encounters with UAP in 2004, they vividly described the object's remarkable maneuverability, and recalled derision from colleagues, including cartoons, jokes, and screenings of *Men in Black* and *Independence Day* on their aircraft carrier. The bureaucratic language of the Pentagon Report had it right: "Sociocultural stigmas and sensor limitations remain obstacles to collecting data on UAP. . . . Reputational risk may keep many observers silent."

Scientists, on the other hand, possess the methodology to assemble the data we need, and where they need better technology to do the searching, they know how to design and build it. The discovery of 'Oumuamua provides the necessary example. When a journalist referred to 'Oumuamua as if it were just another UFO report, I was quick to correct him. This was not another dubious sighting captured by a jittery camera on a speeding military aircraft or a frightened witness alone in a field. The data on 'Oumuamua were obtained and confirmed by scientific observations on state-of-the-art telescopes operated by astronomers equipped with deep knowledge of the normal run of extraterrestrial objects and how they behave—asteroids, comets, and so on. Yet these astronomers had not been looking for UAP. A deliberate search is far more likely to yield the new data we need.

There is urgency. This, too, was clear from the Pentagon Report. Its very existence traces to the US Senate's directive to the Office of the Director of National Intelligence to "submit an intelligence assessment of the threat posed by unidentified aerial phenomena (UAP) and the progress the Department of Defense Unidentified Aerial Phenomena Task Force (UAPTF) has made in understanding this threat." What the Pentagon is built for, and what United States Navy pilots are trained for, is preparation for a threat with the expectation of neutralizing it. The military's job is to assume an unknown phenomenon is adversarial.

The scientists' job is the opposite. Our job is to assume nothing. Science rests on reproducible results that can be replicated by arranging similar circumstances over and over again. The nature of credible scientific evidence is particularly critical in the case of UAP, since skepticism about extraterrestrial technology runs so high. In the wake of 'Oumuamua and the Pentagon's report, the community of scientists should snap back to its natural instincts. These extraordinary incidents present us with a wealth of what we are trained to explore—namely, anomalies. We may or may not find objects built with unusual bolts of extraterrestrial origin or bedecked with the brand logos of alien manufacturers. But we certainly can seek and investigate anomalies of behavior—for instance, motion at unprecedented speeds or accelerations inaccessible to either natural or human-made phenomena. We can analyze signs of intelligent activity—say, evidence of beings who are illuminating or warming the dark side of their planet, seeking or receiving information or responding to circumstances in technological ways that cannot be mimicked by natural processes. In our daily lives, we routinely assess behavioral traits in others to recognize high or low intelligence even before we speak with them. Likewise, we can cite combinations of anomalous physical and behavioral characteristics to establish the case for extraterrestrial technological equipment beyond a reasonable doubt.

There is another reason scientists must take up this task. They cannot avoid it. Inevitably, it will be scientists who conduct the pursuit of knowledge about UAP and ETC. No one else is better equipped to do so, and many pockets of human civilization are less trained to do so. A few could derail the work altogether. That is why the scientific benchmark of transparency is all the more urgently needed in our pursuit of data and explanations. There has never been any place for realpolitik in science for the simple reason that the physical laws governing the Universe pay zero attention to the political gamesmanship of sentient intelligence, human or not. Sadly, this hasn't proven true among scientists, who chase funding, reputation, and the sorts of prizes that get mentioned in the first paragraph of a recently deceased scientist's obituary. Given the civilization-altering potentials in the pursuit of answers about UAP and ETC, we must do our work in the sunlight—a prospect that frightens many of my fellow investigators.

Scientists know that all science is work in progress. We understand that we cannot trap "truth" in amber. We are always in search of more data, and if we aren't, more data will inevitably crash into our awareness to wreck premature declarations of certitude. Yet, even knowing this, many scientists still advise each other to keep the process of exploration behind closed doors. Never show how the sausage is made, they say, or the public will lose trust in the solid scientific consensus about a problem like global climate change.

My view is exactly the opposite.

To succeed in the quest to understand UAP, we need the public's trust. We will deserve that trust only if, step by step, we share the painstaking process by which scientists actually make progress. We must show that where evidence is inconclusive, scientists develop multiple and often competing interpretations. We argue, we seek more evidence, we shape new interpretations to explain what we find. That is how we extend our understanding, not a sign that scientists can't be trusted since they never seem to agree. If more and more people grasp the essence of the scientific method, we will find a rising tide beneath our boats.

If, on the other hand, we engage the public only from a stage at a press conference where "final" conclusions are announced, we will confirm the public's suspicion that scientists are just another self-serving elite, speaking down to ordinary people from a bastion of superiority. And they will quite properly withhold their support. To win their trust, we'll be wise to invoke the spirit of the Nobel Prize–winning physicist Richard Feynman, who once remarked: "Science is the belief in the ignorance of experts. When someone says 'science teaches such and such,' he is using the word incorrectly. Science doesn't teach it. Experience teaches it." That's what we must share with the public, worldwide—our search for observed experience.

Where this search touches on extraterrestrial civilizations and artifacts, it has the potential to change everything. It is already starting to. It is not just a new vocabulary—flying saucers become unidentified flying objects become unexplained aerial phenomena become all-domain anomalies. It is a new frame for understanding our place in the Universe, and more importantly, our purpose as a sentient intelligence on the verge of its interstellar future. Human civilization's science-led search for data on the existence of UAP and ETC is also our search for answers to the most fundamental questions of our existence.

UNSOLVED PUZZLES AND PROFOUND QUESTIONS

Here is another message of *Interstellar*: In our interstellar future, science cannot remain a narrowly bound discipline. Through our experience, science must be allowed to help us define our civilization's grandest ambitions.

This was brought home to me in September 2021 when I received an email from Rob Dobrusin, Rabbi Emeritus at the Beth Israel Congregation in Ann Arbor, Michigan. He wanted to let me know that during the Jewish High Holidays he had given a sermon about *Extraterrestrial*. I read his words quickly—the sermon concerned a phrase from the Rosh Hashanah musaf service, "This day, all the creatures of the universe stand in judgement before You, O God"—before replying that I was humbled and grateful, and added that my mother and father, now deceased, would have been delighted by it. They had been observant Jews who had raised me and my sisters on an almond farm in Beit Hanan, Israel.

In short order, I received a series of emails, from colleagues and strangers, all referencing Rabbi Dobrusin's sermon. One, from a Harvard historian, observed that finding my consideration of 'Oumuamua's extraterrestrial origins showing up in a religious service was striking, "but not surprising given how it bleeds into questions of the meaning of life and humanity's place in the universe."

I went back and re-read Dobrusin's sermon more closely. In it, he allowed that "there is nothing in traditional Jewish faith that would in any way be threatened by assuming or even proving the presence of intelligent life elsewhere in the Universe. Perhaps some of our texts even presuppose this reality." Not only is there no threat to faith, but, Dobrusin wrote, he will "take this one step further. I think this search is vital and can be extraordinarily meaningful on a spiritual level."

I believe Dobrusin is presenting us at the start of our interstellar future with the frame of understanding that must define the purpose of that very future. The search for UAP and ETC has the power to disrupt our hubristic complacency precisely because it holds out new promise of humanity's ability to answer sentient intelligence's most fundamental puzzles:

What is the meaning of our life? If other sentient, technologically sophisticated actors had been around for a larger fraction of the past 13.8 billion years since the Big Bang than humanity, they may have acquired a better perspective about the meaning of life. It is presumptuous of us to grasp this meaning based on the less than 5,000 years of our recorded history. If our own life was seeded by others, this connection would give it a whole new meaning similar to orphans who acquire new knowledge about their parents.

Does God exist? If we mean an entity that can create life or new universes and whose scientific understanding of biology and quantum gravity is well ahead of ours, then discovering evidence of extraterrestrial technology would provide an answer, especially if they possessed abilities that our religious texts assigned to a divine entity.

Is there life after death? Extraterrestrials might teach us how to extend our life expectancy by orders of magnitude beyond what we, with our current medical science and still limited technologies, have accomplished. If death can be postponed long enough, then this question loses its urgency, perhaps even its meaning.

How should humans treat each other? Throughout human history, we have done often horrific things to each other on the basis of minor genetic variations. The realization that there is a far more advanced species out there will make such differences less important, perhaps even unimportant, and could convince us to treat each other as equal members of the same species. We could go one step further: sentient intelligence throughout the Universe should treat all other sentient intelligence as equals.

What should be our goals? A broader perspective provided by confronting the realities of existence and persistence of life far from Earth could, indeed should, reshape our goals to aspire to measure up to the full cosmic context.

We could answer unsolved scientific puzzles, such as: What happened before the Big Bang? What are dark matter and dark energy? What happens inside a black hole? If extraterrestrial science is far more advanced than ours, we might, like a struggling student suddenly gifted the solutions to a difficult exam, learn the answers to many of our unsolved questions. Benefitting from the knowledge of others will also teach us modesty. Albert Einstein, once placed alongside the most impressive Universe-wide intelligence, may not prove to have been the smartest scientist since the Big Bang.

In other words, the stakes in the outcome of humanity's scientific curiosity into the existence of extraterrestrials, and the possible discovery of extraterrestrial artifacts, couldn't be higher. The Pentagon Report was correct to present the implications of UAP in Earth's atmosphere as threatening. But that report misunderstands the threat. A given is that life on Earth, astrophysically speaking, eventually confronts the end of everything. Eventually the hydrogen that fuels the Sun will run out, our yellow star will expand, growing in size to become a red giant that will consume the Earth. If human civilization has managed to survive the 5 billion years it will take before the hydrogen runs out, and if that civilization never summons the curiosity or ability to get off-planet, to travel beyond the Solar system, then whatever that civilization has accomplished to that point will be mooted. Whatever purpose we pursue exclusively on Earth is destined to end with the planet. The same will prove true if, well within 5 billion years, human civilization behaves in ways that hasten its demise, even extermination.

Therein lies the greatest irony. To date it has been humanity's military branches, its professionals trained in the practices of combat and war, who have done the most to advance our evidence of UAP. Unsurprisingly, almost entirely missed by the Pentagon Report is the fact that the search for UAP and ETC is entwined in the hope of preserving all terrestrial life, of advancing not a particular nation state but all human civilization. We'll never meet our interstellar neighbors if we blunder our way into early extinction.

The fastest way to ascend the ladder of civilizations is to reach in hopeful expectation of another civilization (even if only by way of a long-discarded artifact) extending us a hand up. What we have accomplished in our 5,000 years of recorded history tells us nothing about the possible accomplishments of a civilization a million, several million, a billion years old. A certain way to ascend no higher is to refuse to try.

Yes, there are merits in scientists seeking the familiar. If our civilization is capable of producing a technology or a signature of our existence, by transitive logic it is possible another civilization is, or was, capable of the same. Seeking only the familiar is to put on blinders. The only boundaries we have that can guide our search are the physical laws, written in the language of math. These, we can state with high confidence, apply equally everywhere and across time.

In this vein, we should search without prejudice. We should search for the familiar, and the unfamiliar that comports to the physical laws as we understand them. And starting with our generation, we need to begin, with scientific rigor, the search for unfamiliar objects in space that could represent technological equipment sent out by civilizations that predated us in the Milky Way.

In light of all of this, what, exactly, should our scientists be looking for?

The Pentagon Report points us toward only one phenomenon demanding scientific attention: UAP. In the language of the expanded purview of the All-domain Anomaly Resolution Office, the Department of Defense and the Office of the Director of National Intelligence have tasked it with surveillance, reporting, mitigation, and defeat of "unidentified space, airborne, submerged, and transmedium objects." Unsurprisingly, the Office's focus of attention is restricted geographically to military installations, operating and training areas, and special-use airspace. This is too limited a scope of interest, and scientists must do what the nations' militaries are poorly designed to do. We must commence a planet-wide search for curiosities. When it comes to UAP, we must build with transparency and disperse observatories uniquely conceived to gather data that permit humanity to routinely identify UAP such that they are ever more accurately winnowed down into bins of the explainable.

'Oumuamua's tantalizingly unresolved mystery—Was it a naturally occurring phenomenon or a manufactured one?—underscores a task not addressed by the Pentagon Report. Human science needs to become exponentially more curious about near-Earth objects of potential extraterrestrial manufacture. In the vast array of observatories and spacecraft ever created by humans, we currently have none explicitly built for this purpose. The James Webb Space Telescope now allows us to peer ever farther out into space and ever further back into cosmological time. NASA has built and deployed craft to land on a comet for the purposes of mining it and to strike an asteroid for the purposes of diverting its trajectory. What NASA hasn't yet built is the observatory capable of mapping small near-Earth objects, identifying their qualities such that the most outlier objects suggesting a manufactured origin are noted. Shortly, however, the Vera C. Rubin Observatory in Chile will go live and it could be used for that purpose. Very soon, we will be able to track the next several 'Oumuamua. What we don't have is the means to intercept, image, or capture the next 'Oumuamua.

The least expensive exploration for extraterrestrial objects is that which can be done without leaving Earth. We know to a certainty that interstellar meteors strike the planet, and we now know to a certainty where the remnants of two, IM1 and IM2, rest. Only one in a thousand meter-sized meteors originate from interstellar space, so interstellar meteors like IM1 and IM2 are difficult to sort out from the collection of all space rocks, without prior knowledge of their high speed. In this regard, the relative youth of our Solar system is a possible plus. Having arrived after over 9 billion years of cosmic history had already past, its planets, including Earth, could well have become repositories of discarded materials from far older civilizations. The nascent academic field of interstellar archaeology calls out for support and funding.

Within a matter of a few decades, humanity has sent five craft—Voyagers 1 and 2, Pioneers 10 and 11, and New Horizons—already in or destined to enter interstellar space. These chemically propelled rockets will quite possibly be passing through that space millennia after humans have managed their civilization and planetary resources poorly enough to effect their, and even terrestrial life's, extinction. But, the longer we persist, the more often we are likely to send craft out into interstellar space. And the opposite logic holds true: any civilization similar to ours that managed to last for millions of years could well have sent out billions of such craft. It is high time scientists looked deliberately for them.

Our new vocabulary needs to pivot from what we're *least* likely to encounter—a craft operated by a biological creature—to what we're more likely to discover—an inanimate object or a sentient one.

In much of the science fiction we love, from the novels of Ray

Bradbury to the films of George Lucas, we see biological beings, human or otherwise, crossing the unfathomable distances between stars, even between galaxies. But the conjurers of these fanciful voyages sidestep a regrettable fact: Darwinian evolution has not selected biological beings to survive interstellar travel. Even light, traveling at 186,000 miles per second, requires tens of thousands of years to travel between stars across the full extent of the Milky Way disk and ten times longer across its halo. A long trip, by interstellar standards, must span a great many human generations.

This means that we are far less likely to encounter biological beings in space than we are to find technological equipment that biological beings manufactured. If the equipment was launched long ago, it may now be worn out or broken. Imagine our own Voyager spacecraft a billion years from now; it's not going to look like it did the day it came off the showroom floor. Likewise, humans produce far more waste products than they do deliberately built, well-calibrated scientific instruments. Like plastic bottles in our planet's oceans, it's easier to imagine that technological debris more akin to extraterrestrial trash from distant civilizations may have been accumulating in interstellar space for billions of years.

Some of it could well have landed on the planets of the Solar system, and those planets with little to no atmosphere may well hold the most intact deposits. Again, we will not know until we go and look.

Whether the technology out there is in good repair or broken down, we should expect it to be carrying artificial intelligence (AI) systems. To save energy, such systems might shut down during long journeys, then switch on when they get close to stars, using starlight to recharge energy supplies. They might also turn on with tampering or even in response to being observed. To manage encounters with extraterrestrial AI, we will likely need advanced AI of our own. Such systems may be needed to properly interpret the signals and behavior of alien AI.

Scientists should also look backward in time, seeking and pondering evidence of anomalous events in the long history of Earth.

For example, 2 billion years after the planet was formed, the level of oxygen in the atmosphere rose sharply, enabling the emergence of complex life forms. Why the rise? We don't know. Nor do we understand why intelligent life abruptly appeared only in the last one-thousandth of the planet's history. Both events probably have a terrestrial explanation. But extraterrestrial causes are possible. Regardless, knowing the biological mechanics of life on Earth will prepare us for considering the knowledge our own spacearks must carry.

All of this scientific work of a very youthful interstellar civilization will advance more quickly, more significantly if we also start the work of fundamentally shifting our sense of our place in the Universe. We have spent centuries in historical stages of preening self-regard. Master races and superpowers are the limited planetary projections of a transparently egotistical, cosmically late-blooming sentient intelligence. The longer we allow that vocabulary, that frame of understanding to determine our future paths, the longer we will wait to enter by craft and astronaut our interstellar future. And the greater the likelihood we fail to enjoy any such future at all.

Our scientists must begin immediately the work of gathering new, more, better data. All humans should petition their leaders for greater transparency as to the data already held by governments, but still kept secret. We need ever more leaders worried about human prospects alongside narrow, nationalist constituencies. I was reminded of this in 2021 when John Ratcliffe, the former Director of National Intelligence, told a reporter for Fox News: "We are talking about objects that have been seen by navy or air force pilots, or have been picked up by satellite imagery, that frankly engage in actions that are difficult to explain, movements that are hard to replicate, that we don't have the technology for."

Ratcliffe may have inadvertently revealed that even more evidence exists than we know. The federal government has never released images of UAP captured by satellites. If such images exist, the scientists pursuing evidence of extraterrestrial life would very much like to examine them, looking for data on objects that enter the Earth's atmosphere but do not follow ballistic orbits like meteors. Such data may become available as the new federal office charged with coordinating efforts to report and respond to UAP gets up and running. The office may determine that the objects Ratcliffe mentioned are so unusual they cannot be human-made and hence are not a matter of national security. If so, then the data should be offered to scientists for analysis because it is of interest to the entirety of humanity without adhering to national borders.

In short, along with building new data-gathering instruments, scientists must also go treasure hunting among released, previously classified archives. Because of classified data released by the United States government, we now know that 'Oumuamua was not the first detected interstellar object. It is only because of that reluctantly released data that we have within a year identified a second interstellar meteor. Ahead of the Pentagon Report's release, Ratcliffe allowed what is tacitly understood by public and scientists: "Frankly, there are a lot more sightings than have been made public. Some of those have been declassified."

Pondering the possibility of extraterrestrial life, the great physicist Enrico Fermi once asked: If conditions favorable to life are so common in other solar systems, then "Where is everybody?" Why have we had no visitors? Why don't we see life forms strolling out on the galactic streets? But this fails to reckon with the lifespan of the Universe, with its boundless stretches of time over billions of years. The age of the Milky Way is a million times greater than recorded human history. The visitors may have been here. In addition, only over the past decade had we developed instruments that allow us to detect the first interstellar objects. And the first three among them, namely 'Oumuamua, IM1, and IM2, appear to be outliers in their physical properties relative to the familiar population of space rocks from the Solar system. In the past, we may just have missed technological visitors. But they may have left traces that a Sherlock Holmes of science—or, better yet, a generation of such detective-scientists—can find.

2

THE DAWN OF OUR INTERSTELLAR FUTURE

HUMANITY'S INTEREST IN THE stars transcends culture and time. Long before recorded history, our earliest ancestors looked into the night sky and onto that great, bright, swirling chaos of objects they began to impose order. They saw patterns in the stars that became constellations that represented gods and heroes and sacred beasts. They weaved stories about these patterns into their myths and folklore. Though these stories varied tremendously across Earth's civilizations, all stargazers looked on a common Universe and noticed that different celestial objects moved in different but predictable ways over time. And, over time, the stories about the stars evolved, as did our understanding of our relationship to them. Were they pinpricks in a curtain, obscuring heavenly light? Were they distant suns occupying different celestial spheres? Did they contain divine messages for humans to interpret? Were we at the center of it all, or just passengers on one particular planet, a small part of this great whirling mass? Answers varied, but the curiosity (and with it the pursuit of better answers) was held by all.

Babylonian astronomers were among the first to differentiate planets from stars. The Chinese were among the first to document a supernova. Persians spied the Andromeda Galaxy first, and the

Greeks grasped that the Earth was a sphere. For human stargazers on every continent, the movements of the sky became ways to orient themselves in place and time, and they built elaborate buildings and monuments to study them. They were inspired to devise calendars and writing instruments to record and pass down knowledge and wisdom. And when civilizations crossed paths, peacefully or not, some of that knowledge, some of that wisdom became a joint inheritance. The sky and, above it, the expanse of space was humanity's first shared data set, and one of our first universally shared interests.

Over millennia myths gave way to ever more confident, robustly reconfirmed hypotheses. A rising wealth of data available to all human scientists laid to rest some questions and identified many more still in need of hypotheses, data, and confirmation. Over the last several hundred years, observation and scientific analysis, expressed through degrees of mathematical certainty, have become our shared language of exploration. But our separate civilizations' instinct to look into the night sky and weave unique tales about what we've seen remains strong.

The desire to plant nation-state flags on planets, for example, remains powerful. That the governments of Earth behave this way is unsurprising. That human science needs to collaborate with, but not become corrupted or corralled by, governmental concerns is a truism of our early interstellar history. This is why, on July 26, 2021, a little more than a month after the Pentagon Report on UAP became public, the Galileo Project was announced to the world.

The two closely timed events are related only as examples of our having begun the first years of humanity's interstellar era. The United States government and its military had collected, mostly clandestinely, anecdotal data on UAP for decades before the Pentagon issued its report. The Galileo Project is a response to the one uncontestable fact 'Oumuamua's passage left: humanity's gap in instrumentation and data necessary to answer if near-Earth extraterrestrial objects exist. Months before the Pentagon admitted its lack of useful data concerning UAP, the Galileo Project was being conceptualized. Which is why a month after the Pentagon went public, the Project could declare its goal, "to bring the search for extraterrestrial technological signatures from accidental or anecdotal observations and legends to the mainstream of transparent, validated and systematic scientific research."

What would it take? The origin story of the Galileo Project doesn't start with curiosity about what might be up there, which is thousands of years old, but with the more practical questions regarding the instruments, staff, and funding it would take to learn if extraterrestrial artifacts were in our Solar system. First one, the passionate scientist, Harvard visiting scholar, and biotech entrepreneur Frank Laukien, and then ever more stargazers reached out to say that not knowing if 'Oumuamua was or wasn't of extraterrestrial manufacture is unacceptable. What would it take to scientifically answer the question, "Have other extraterrestrial civilizations left discoverable artifacts near Earth?" I wrote a white paper listing what it would take, funders came forward, and I agreed to lead the Project, which would be multi-institutional and international, with its home being the Harvard College Observatory. For almost as long as humans have had a recorded history, we have expressed our awe and awareness of the Universe, but even so, this project represented something new, a research group devoted to identifying the nature of UAP and 'Oumuamua-like interstellar objects by using the standard scientific method and a transparent analysis of open scientific data.

The Project dedicated itself to three major avenues of research.

Our goal is to obtain high-resolution, multi-spectral UAP images, with the intention of discovering their nature. The long-term ambition is to build and maintain a global network of mid-sized, high-resolution telescopes and detector arrays with suitable infrared, optical, radio sensors as well as computer systems—equipped by artificial intelligence (AI) algorithms to distinguish between natural objects like insects or birds, human-made objects like weather balloons, near-space surveillance craft, drones, airplanes, or satellites, and something unknown. Immediately, the Project focused on building the first such UAP-specific observatory and debugging it atop the roof of the Harvard College Observatory. After the initial test period, the suite of instruments was moved to another part of Harvard and several copies of it are now being assembled in additional locations.

The Project's second aim is to begin the work of conceptualizing and designing a launch-ready space mission to image unusual interstellar objects by intercepting their trajectories. To do this we are collaborating with other space agencies and space ventures and utilizing existing and future astronomical surveys, especially the Legacy Survey of Space and Time (LSST) at the Vera C. Rubin Observatory, currently nearing completion in Chile. Finally, using advanced algorithmic, AI object recognition and fast-filtering methods, the Project will discover and catalog satellites that are exploring Earth. That most, perhaps all, such satellites will be of human manufacture will be knowable only once scientists have access to such a catalog. It is also the means of discovering the one, or several, not made by humans.

Within months of its launch, the Project added a third branch of inquiry to its agenda: the search and recovery of anomalous interstellar meteors, or the first explicitly interstellar archaeological dig. The Project, like the Department of Defense's Anomaly Resolution Office, is now investigating all domains. What is in our Solar system? What is in our atmosphere? What is under our oceans? And what evidence is there that any anomaly discovered suggests an artifact of extraterrestrial manufacture?

As important as the goals of the Project are its principles. It is to be a transparent scientific project to advance a systematic experimental search for cross-validated evidence of potential astro-archaeological artifacts or active technical equipment made by putative existing or extinct ETC. That its foundations are deeply grounded in the scientific method makes all the difference. The Project is clear that its animating interest is in scientifically addressing the possible existence of extraterrestrial artifacts and civilizations. It is equally clear that its results could simply be the gathering of rich data sets that discover and scientifically explain novel interstellar objects with anomalous properties, previously unexplained but natural phenomena, and terrestrial technology currently cataloged as inexplicable UAP.

Within months, the Project grew to include twelve members of a scientific advisory board and over one hundred volunteer research affiliates spanning areas of expertise from astronomy and astrobiology to machine learning and artificial intelligence (AI) algorithms. They were tasked with thinking through the practicalities of gathering the desired data: How to construct the hardware and software platform to localize UAP. What equipment was needed to monitor multi-band acoustic aerial signatures. What will be required for a rendezvous mission with ʻOumuamua-like interstellar objects given the primary objective of a clear, resolved image of that object. This required careful consideration of how small an object can be reasonably targeted, and what information is needed to more accurately expand the current population of reachable interstellar objects. And, with the discovery of the first two interstellar meteors, IM1 and IM2, it required identifying the necessary research vessel and underwater equipment, and designing and constructing the recovery technology necessary to retrieve interstellar meteorites on the ocean floor.

Electrical and software engineers joined with astronomers, physicists, biologists, chemists, and computer scientists to blaze new trails. From the outset, the Project adopted a big-tent approach. Our ranks included the skeptical and the convinced. Among the former, Edwin Turner of Princeton University would provide the most robust case for the probability that human civilization is in fact likely alone in the Universe and that the search for extraterrestrial artifacts is consequently nearly guaranteed to be an unrequited one. Turner kept company with, among a great many others, Luis Elizondo, a former counterintelligence special agent and the former director of the United States' Advanced Aerospace Threat

Identification Program (AATIP), who resigned from public office citing the need for more public exposure and government cooperation in the study of UAP.

Attracting interest and volunteers was not difficult. Further evidence of humanity entering its interstellar future was the fact that the ranks of volunteers interested in helping was deep and talented. Organizing the Project, and directing it, was only slightly more complicated. While the week-over-week administration of the multipronged project would require hours of Zoom meetings, detailed reports, and spirited debate from opposing points of view, we never lacked for a guiding compass with a commonly agreed-on true north. Every project participant agreed to adhere to four ground rules.

First, we would work only with scientific data openly available for peer review and gathered by our own observations. Classified government information was to be off-limits since it could not be shared openly with the scientific community. Same with earlier data on UAP. We could stand by our findings only if they were based on observations for which we ourselves could vouch—and show our work.

Second, we would analyze our data using only the rules and tools of known physics. The Project's scope would remain in the realm of scientific hypotheses, testable through rigorous collection and analysis.

Third, we would share our findings freely and only through traditional channels of scientific publication that adhere to the process of peer review. That means subjecting our findings to the informed, critical, and experienced eyes of scientists not involved in our project.

Finally, we would release no results except through scientifically accepted channels of publication. Our work is for the purpose of scientific discovery, not public relations.

It is in this spirit that I have often observed to colleagues both within and outside of the Galileo Project that there is no such thing as a scientific breakthrough that complements our ego. No small number of Harvard undergraduates and graduate students have heard me observe that the fruits nature has to offer do not fall into humanity's lap because we deserve them. We have to select our path, and then find the fruit where it hangs. Harvesting it is a matter of either being among the first to encounter a particular branch of scientific inquiry or benefiting from years of others harvesting a particular orchard of scientific research. In the first instance, there are the overlooked and unexplored opportunities, reached just by thoughtfully being at the right place and time. In the second, to reach those higher, cultivated branches, you must stand on the shoulders of others.

A clearly demarcating line between a civilization embarking on its interstellar future and a civilization long familiar with its interstellar history is that the latter understands the search and discovery of extraterrestrial civilizations is a commonplace pursuit. The fact that much of humanity still considers the existence of UAP and the possibility of extraterrestrial artifacts, whether passing through the Solar system or resting on the ocean floor, titillating is a sign of our civilization's interstellar immaturity.

Do not misunderstand me. The prospect of discovering an ETC, of establishing to a certainty that a single UAP is of extraterrestrial manufacture, of locating part of an ancient artifact not of human design, is heart-racing. And I can think of no scientific discovery that will more fundamentally transform our civilization and brighten its prospects than proof we are not the sole technologically capable civilization in the Universe. The sooner the search for UAP and ETC joins the disciplines of accepted scientific inquiry, however, the better. Initially, Darwinian natural selection was a matter of furious debate; now it is the stuff of introductory biology. The discovery of the first correctly identified Neanderthal fossil in 1856 resulted in a period of public attention, study, and debate, only to be followed by decades of ever more specialized, scientific, and less publicly attended debate. So, too, should be the history of our discovery of other ETC. And the sooner we arrive at a scientific consideration of UAP and ETC, the better. We have spent decades stigmatizing scientific interest in extraterrestrials, which is in part why humanity has spent far more funding entertainments around their possible existence than in the systematic search for evidence of their existence. This is also why we must place the efforts of the Galileo Project in the context of equivalent scientific pursuits.

Two equally true statements based on data and theory are: dark matter of a particular type might exist; extraterrestrial civilizations might exist. Both involve the search for material entities that are not found in the Solar system. A yardstick for how far into our interstellar future we have traveled will be how equivalent are humanity's efforts to discover empirical proof that either, or both, exist in fact.

It is material that the Galileo Project is not the only scientific project I direct. I am also the director of the Institute for Theory and Computation and the founding director of the Black Hole Initiative, both, like the Galileo Project, at Harvard University. In addition, I chair the Breakthrough Starshot Advisory Committee, part of the Breakthrough Initiatives program established by Yuri Milner, the Russian-born Israeli physicist, venture capitalist, and entrepreneur. All of these projects blend astrophysical theory, observational evidence, and technological competence for the purpose of advancing what we know about the Universe, the physical world and the matter it is made of, and the laws that govern it.

It is plausible that life and technological civilizations are common throughout the Universe. It is also certain that, if discovered, dark matter is far more common. For that reason, discovery of the latter is significant to our pursuit of evidence of the former. Dark matter will be among the most common features of existence that we will have in common. It is far more likely that when human civilization first crosses paths with an extant extraterrestrial civilization, or its AI, the conversation will not be led by the two civilizations' poets. That task will fall to the scientists. (The likely nature of our first communication with an ETC is another thing science fiction has poorly prepared us for. That exchange is far more likely to consist of the sharing of scientific notes and notations rather than military conflict.) Because science generally, and physics particularly, is going to be the lingua franca of universal sentient intelligence, dark matter is among the terms our new interstellar vocabulary must include for the simple reason that it is likely to be one of the shared subjects we will discuss with an extraterrestrial intelligence.

Rather than anticipate having to respond to the directive, "Take me to your leader," we need to prepare to answer a more likely and more interesting request, "Tell us something we don't know."

THE SHARED LANGUAGE OF SCIENCE

One morning in early April 2022, I read a preprint of a forthcoming scientific paper making a strong case for the existence of self-interacting dark matter based in the most recent data on dwarf galaxies within the Milky Way. This led me to a 5:15 a.m. hypothesis that became a 2:00 p.m. paper, "Effective Self-Interaction of Dark Matter from Gravitational Scattering," which was accepted for publication in *The Astrophysical Journal Letters* within a week. It was the shortest "from farm to table" work of my scientific career and was based on the simple idea that the dark matter is organized into clumps, each weighing ten thousand Suns—so that the needed self-interaction is communicated by the single force we are sure the dark matter respects: gravity! And within a few hours of the preprint of my article becoming available online I received a cordial email from another astrophysicist, Jo Bovy of the University of Toronto, casting doubt on my hypothesis and proposing a clean test to set the issues to rest based on future data analysis concerning thin streams of stars in the halo of the Milky Way. I replied with scientific agnosticism: "I would love to see the results."

There, distilled to a paragraph, is the scientific method confronting an unexplained mystery. In this case, just what is dark matter? That we do not know is an embarrassment, for it (ignoring dark energy) makes up five-sixths of the matter content of the Universe. We have spent centuries observing and learning about observable matter, such as comets, asteroids, planets, and interstellar gas. As a result, we know a great deal about one-sixth of the cosmic matter.

The vast majority of what makes up the matter of the Universe, however, we're close to clueless about. Close to because over ninety years ago we were entirely clueless. Back then we simply thought the one-sixth we knew a good bit about was all the matter there was. Though blissful, our ignorance tends to be the first incontrovertible discovery that follows our encountering anomalous data.

In 1933, the astrophysicist Fritz Zwicky discovered dark matter. Discovered is accurate, dark not so much. Zwicky was studying the Coma galaxy cluster. More specifically, he was studying how fast galaxies are moving in it. Given their characteristic speed, these galaxies could only be bound gravitationally if the cluster had invisible matter.

If you have ever sat on a playground sit-and-spin you'll get the idea. One fifty-pound child sitting near the very center of the sit-and-spin doesn't spin very fast. Six fifty-pound children sitting at either the middle or toward the edge of the sit-and-spin can cause it to spin much faster. In short, its speed of revolution depends on the weight of the children and where they're sitting. The same goes for the matter making up the Coma galaxy cluster.

The discovery of dark matter in this and other contexts was like seeing a sit-and-spin rotating as if it had thirty-six children on it, but only six children were visible. The missing mass became the foundation of the simplest of hypotheses: there is something out there that we do not understand. Zwicky called this invisible matter "dark matter," though in truth the term was a placeholder for "another unexplained scientific anomaly."

To humanity's great credit, scientists didn't shy from the embarrassment of Zwicky's data. Before his discovery, we were confident of our increasing grasp of the matter of the Universe; after it, we understood we were ignorant about the vast majority of the cosmic matter. And so, over the past eight decades many of our best minds, arising in numerous nations and working across continents, have presented hypotheses and data that would explain the nature of dark matter.

While it is unlikely that modern human science, a bit over a century old, will tease out insights that would impress an extraterrestrial civilization a million years old, it is not impossible. If any do or did exist, and if any reached or surpassed our technological-scientific prowess, we know for a certainty that their and our scientific efforts are at work in the same orchards of curiosity. Like early humans staring up at the cosmos, other civilizations stare at the same Universe. No more than human civilization can extraterrestrial civilizations lay proprietary claim to any discovery about the Universe, its matter, and its laws. This tells us that while the means of interstellar species communication may be wildly unfamiliar, one subject—astronomy—will be shared. All civilizations that construct instruments capable of estimating the mass of galaxies will see what humanity sees through its own. So, while I am 99% certain they will not refer to five-sixths of the matter in the Universe as dark, I am equally certain that evidentiary data about that matter will be shared, though one civilization may enjoy a far greater trove of data. Differences will be due to the sophistication of their instruments, their tools of harvest, and scientific erudition. Those differences will not be traceable to the common cosmological mysteries we confront but to our separate cultures of science.

Among the challenges our youthful interstellar humanity confronts is that the culture of terrestrial science as captured by its history is, kindly put, a herky-jerky affair. The reason is entirely traceable to humans. Recall that the laws of physics are universal throughout the Universe and across time. The implications of this are many and include the fact that airplanes could have flown centuries earlier than the Wright brothers' acclaimed twelve-second-flight in December 1903. The brothers rightly celebrated that accomplishment. Humanity, however, should have rent its hair. For there is no good reason it took so long for someone to figure out thrust, lift, weight, and drag. And many bad reasons for why it did. Here is one: for centuries women were discouraged, often outright prevented, from being trained and employed as scientists.

Where humanity is on the ladder of civilizations is entirely traceable to the efforts, and self-imposed limits, of humanity.

We must bring humility to the appreciation that we are quite likely below the median of all technological civilizations in the Universe. As a statistical matter, it is highly probable that terrestrial humanity is just one civilization among many and, both as a matter of statistics and a matter of the honest appraisal of our history, not a very impressive one. Of course, among the many things we do not know is just how unimpressive. For centuries, however, the leading edge of scientific inquiry defaulted to the belief that subsets of the species—men, and for much of the West, white Christian men—were uniquely impressive. This was prima facie untrue even at the time it was asserted—consider just the debts European science owed Arabic science as the former emerged from the Middle Ages. It has also indisputably retarded humanity's progress. Likewise, we know for a certainty that we are now a civilization in danger of self-destruction, with much to learn. So, our hope must be that there are more advanced civilizations, and that some of their artifacts exist relatively nearby. Our hope must be that other intelligent life are, or have been, well above us on that cosmic ladder.

This hope frames our generation's single most profound scientific question and ambition. We must seek answers to a straightforward question: How might we accelerate our ascent up the ladder of civilizations to see what those that have come before us have to teach us? Whether by design or not, ETC may well be offering a helpful hand to humanity. There could be artifacts left intentionally for our discovery, or unintentionally. Earth could have been, might be, or could become a deliberate or accidental stop for an interstellar traveler. This frame asks of us at minimum this optimistic practice: rather than presume that all we know of civilized life in the Universe, humanity, is all there is to know, we need to scientifically hypothesize that intelligent life existed before us and, humble and hopeful, scientifically seek any discoverable help.

STUDY IN SCIENCE

What encouraged me to avidly step into the leadership of the Galileo Project was, first, my enduring interest in science's ability to increase the store of humanity's useful knowledge. And second, the prospect that by a single discovery we might advance not a partial way up the ladder of civilizations, but whole rungs at a time.

That possibility begins with a far slower, deliberative effort. What follows is a glimpse into how the early work of science gets done, especially at the start of a field of discovery. This lifting up the hood on the scientific process is, I think, far too rarely done. To the casual onlooker skimming the headlines, the federal report on UAP and the brief transit of 'Oumuamua through our Solar system offered tantalizing possibilities. For most people, the typical and understandable response was to raise an eyebrow, perhaps feel a shiver of awe or fear, and then to move on with the business of the day. These objects *might* constitute evidence of extraterrestrial civilizations. These objects *might* constitute evidence of terrestrial espionage deserving of a Sidewinder missile. But what could you do? It was simply one more of life's mysteries. What is inducement for most people to get on with their day is inducement for scientists to get to work, or it should be.

To a scientist a mystery is a challenge. In the language of science, a mystery is an anomaly—a thing that departs from the norm, a thing that cannot readily be placed in a class of the known, named, and understood to consistent predictability. Whether UAP in the sky over New Jersey or a pancake-like object a football field in size passing through the Solar system, these objects unquestionably exist and unquestionably exhibit characteristics we cannot explain. To the scientist, that is—or ought to be—unacceptable. What is more, they should be inducement to act. The scientist regards the anomaly precisely as Sherlock Holmes regarded the unsolved murder. It is an invitation, a provocation, a goad.

Tellingly, when readers are first introduced to Holmes in *A Study in Scarlet* they find him in a research laboratory devising a means to more accurately gather data from blood samples. It is a reason I find the fictional detective so useful an analogy to my day job. From the very start, Holmes's powers of practiced perception were backstopped by better data gathered by the scientific method. In story after story, the average-witted inspectors of Scotland Yard jumped to conclusions. In story after story, even those in which Holmes formed a theory of the crime at the outset, he would declare that more evidence, and more conclusive evidence, was needed. The investigation must broaden so that the pool of suspects narrows.

So it has been with the Galileo Project. We must bring our particular anomalies from the realm of the unknown to the realm of the known by acquiring more evidence. Unlike Holmes, however, our initial work did not begin in the laboratory. Most of the first months of the Galileo Project's history were spent raising funds, attracting talent, and then organizing staff into committees and subcommittees. To a scientific advisory board was added a philanthropic board and public outreach affiliates. Thereafter, there was an ever longer list of research affiliates across numerous disciplines. The first things the members of these committees did was introduce themselves; thereafter, they presented, debated, queried and answered, and every now and then argued or cheered. It is commonly understood that the constraints of what is possible are bounded by knowledge, technology, and finances. Less appreciated is how necessary joy and optimism are, albeit bounded by the scientific method. All of these were factors in that initial work of defining the methodologies and describing the necessary tools the Project would require for the systematic and objective pursuit of useful data.

What, precisely, are our aims? What are we trying to find out? These questions take on additional importance whenever stigmas attach to a scientific inquiry. Going in, we knew from the history of SETI that too many, including too many critics, would declare we were seeking LGM, the sarcastic acronym for Little Green Men. For the UAP branch of the Galileo Project, the opposite was true.

We began by accepting the face value of the term "unidentified aerial phenomena." Our scientific task was to eliminate the "un-." We would test what scientists call a null hypothesis, which for the UAP branch was: "All aerial and astronomical observations can be explained as known phenomena." Perhaps our new data would prove that hypothesis. Perhaps not. A given was that our data would narrow the scope of what remained unexplained.

We knew, of course, that any observatory we built would identify numerous "explainable phenomena"—familiar ones such as airplanes, helicopters, drones, birds, insects, clouds, swamp gas, balloons, satellites, and hoaxes. If US Navy pilots were catching glimpses of aversive technologies operated by America's adversaries, we would likely encounter those, too. We would need instruments and programming that would swiftly identify and analyze all objects that crossed our observatories. This alone would be a valuable scientific pursuit, useful to everyone interested in safely maintaining the skies. Part of our goal, then, was to use the collection of data on UAP to develop and apply state-of-the-art artificial intelligence methods for autonomous detection and characterization of all aerial phenomena, explained and not. In addition, the Project would need to develop a suite of instruments for monitoring the terrestrial atmosphere and by doing so create standards for detecting aerial anomalies. As a result, we would establish a new and reliable scale for reporting such anomalies with various levels of confidence. A secondary consequence might be fewer aversive objects. If observatories capable of identifying all explainable aerial phenomena were deployed across the United States, or the world, deploying clandestine UAP would be mooted.

In a twist on Sherlock Holmes's adage about finding the improbable truth by eliminating the impossible, the Galileo Project's UAP research would eliminate the explainable aerial phenomena of terrestrial origin. If our instruments and computer software succeeded in explaining all such phenomena, then UAP would cease to exist, having transitioned over to EAP, explained aerial phenomena. However, if after narrowing down most UAP to EAP there remained a few of the former, we would, à la Holmes, have eliminated the terrestrial and whatever remained would be ever more plausibly of extraterrestrial origin.

Having defined our aims, the next set of debates drilled down on how to get the data we wanted, what equipment we would need?

A devotee of science fiction might naturally assume the Galileo Project would have to invent some twenty-first-century version of Tom Swift's Megascope Space Prober. In practice, we started with something that to the lay ear might sound as fantastical, the Project's science traceability matrix, or STM. Far from anything fantastic, an STM is a standard element for space exploration proposals that spells out scientific goals and objectives that inform instrument and mission requirements. Not unlike considering just what you will make for dinner before you go to the supermarket, we needed to articulate the guidelines of our theoretical analysis, interpretation, and modeling so that what we bought, built, and coded would accomplish some of what we intended. With this done, our astronomers and engineers were able to in fact write a shopping list of instruments that could be purchased right off the shelf, many at surprisingly reasonable costs.

On that list were just six necessities: infrared and visible-light cameras that take a video of the entire sky; a passive radar that detects the reflection of radio waves from terrestrial broadcasting off objects in the sky; an instrument for surveying particle count, weather, magnetic field, atmospheric opacities, and UV atmospheric glow; audio including infrasound and ultrasound instrumentation; and computer software. It was also understood that no matter the outcome we would be public about our results. Like offerings in a competitive cooking contest, we would be rigorous in our preparation and after all due diligence show the results to critical, experienced judges. Presentations of our findings would go to external panels and peer-reviewed journals.

Where to build the first UAP observatory? The roof of the Harvard College Observatory, and this despite the fact that it was commonly agreed to be a terrible place from which to seek UAP for reasons of light pollution. But it was among the best possible places to debug a prototype. On the roof, it would sit more or less, at least initially, directly over my head when I sit in my office. Placed there, I, and multiple members of the Galileo Project, would be able to walk among its instrumentations with ease.

Those instruments, we expected, would include three fast-slew mounts for different instruments. One, managed by a targeting computer, would support a guiding camera and medium resolution camera. The same computer would be tasked with managing the data coming in from a fisheye camera. A passive radar on one mount would accompany audio instruments on another. A third table would host the telescopes. Early on, the sketches of the observatory instrument that would provide a 360-degree, 24-7, all-weather view of the sky visible to the observatory showed a small dome that looked very similar to the half-spherical top of R2D2, of *Star Wars* fame. Its array of cameras would feed data into a target coordination system. Taken together, to the uninitiated they were anticipated to make the department's rooftop appear like that of a home over-dedicated to television and ham-radio antennae. To the initiated, however, they would look like the first step on the path to establishing just what the reported Tic Tac–shaped objects captured in jerky video footage taken by US Navy pilots are.

From the outset, all the instrumentation and computer hardware was designed to fit well within a cargo container to ease the ability to place and move these UAP observatories to where they would gather the most useful data.

Unfortunately, no one currently knows where those locations are. This is relevant to the design of the observatories—Would they sit in dense forests or deserts, near densely populated towns or far removed from all humans, at sea level or at high altitudes?—and the programming of the governing software. To date, there are no published studies about the frequency of UAP events. Galileo Project scientists necessarily worked from the two available databases, the National UFO Reporting Center based on civilian reports, and another based on the published reports of pilot sightings.

From the National UFO Reporting Center, which has collected reports since 1974, we analyzed 48,531 sightings from 2010 to 2020 sent in from some ten thousand cities and towns. We found unusually high numbers of reports in ten counties in seven states from the east coast to the west. These were arrived at by regression analysis, which, of course, asserts and assumes no causality. It also provides no guidance on why these areas are "hot," or even if, after study, they are in fact. The obvious correlation is of the frequency of sightings with population density. After all, local culture or a nearby military base could explain the rate of civilian sightings. Note however, that uncovering even this advances the null hypothesis, pushing some unexplained phenomena onto the balance sheet of the explained. That admitted, our best probability study, which, again, will be refined over time as more and better data accumulate, points to an initial ambition of each observatory covering a "hot spot" region of ten kilometers on a side. Given current funding, the modeling suggests that with eight observatories operating for five years we can anticipate detecting tens of aerial anomalies. As is always true with terrestrial science, were funding to increase, so too would the extent and quality of the data collected. As would the likelihood of unexpected, perhaps wonderous discoveries.

Early on, the Project agreed to follow the Confidence of Life Detection (CoLD) scale, which is also embraced by NASA. Composed of seven layers, each layer is intended to show how close scientists are to detecting extraterrestrial life. Most relevant for those branches of the Project that were looking out into space for sufficiently interesting interstellar objects, the CoLD scale is useful across all scientific endeavors to identify possible extraterrestrial civilizations and their artifacts. At the lowest scale is detection of a signal from a biological entity. Thereafter, causes for contamination of the data are ruled out, all known non-biological sources of that signal are demonstrated to be implausible, additional independent signals are sought and detected, alternative hypotheses are offered and systematically ruled out through further observations, and, finally, an independent follow-up observation of predicted biological behavior is undertaken.

While the UAP branch of the Project was conceived at a scale that could almost immediately move from debate and presentation to purchases and construction, the Interstellar Object team, or ISO branch of the Project, could not. Like the UAP branch, the ISO branch confronted a paucity of already established data. Right now, humanity is woefully ignorant of the number and nature of interstellar objects passing through the Solar system. While that is going to change soon as ever more sophisticated telescopes and observational equipment are completed and made potentially available, at the Project's inception much of the ISO branch's work would necessarily be theoretical. What that branch knew from the outset was its interest wasn't in all interstellar objects, but rather in the subset of them that presented with properties outlier to all known Solar system originating objects. 'Oumuamua provided them with a useful example of one such object. What instrumentation would be necessary to identify the next 'Oumuamua-like object? And then there would be the need of a spacecraft to rendezvous with that object, and ideally a craft that while expensive was not so prohibitively expensive that planning for several intercept missions was off the table.

THE VALUE OF IMMATURITY

Which takes us to the question of expense. For the Galileo Project to build its first, and to date only, UAP observatory cost about $250,000. The expedition to retrieve the fragments of IM1 on the ocean floor will run about $1.5 million. A cautious projection to establish a mission to rendezvous with the next 'Oumuamua-like object will run over $1 billion. Total spending on the James Webb Space Telescope was estimated to be about $10 billion as of 2022. Some estimate the most expensive object ever constructed and maintained is the International Space Station, at about $160 billion. In 2022, the president submitted a national defense budget request of $715 billion. And, according to the US Bureau of Economic Analysis, in the second quarter of 2022, consumer spending in the United States was roughly $14,000 billion.

The Pentagon Report, but even more so the United States government's decision to fire missiles at UAP in 2023, prepares us for aerial phenomena with potentially bad intentions. Report and missiles reflect the biases of military and government as well as the organizations that gathered, and are gathering, UAP data for them. It reflects decades, very likely centuries, of fearful humans who routinely shared accounts of otherworldly visitors possessed of vast powers but animated by humanity's own most base interests. We have far more often envisioned marauders than we have benevolent educators. Either presumption imagines discovery of an artifact with intent. Most often, we envision something biological, and rarely do we allow that any visitor traversing interstellar space is more likely to be artificial intelligence. Here, again, we need to widen our aperture of scientific interest by transitive logic. Of all the things humans produce, first among them are consumer goods, most of which are single- or brief-use items. A distant second to those items are materials produced for the conduct of defense and aggression. And a distant third to those items are objects with explicit scientific purposes.

It is possible that 'Oumuamua was a signal buoy awaiting an encounter with our Solar system. It is possible that it was an extraterrestrially manufactured dandelion, and encountering our Sun and registering our Earthly civilization triggered the release of micrometeors invisible to our observational tools. It is also possible that it was a lone leaflet, one among millions that another civilization once spread throughout its corner of the Universe, a pitch for craft to use this particular service station over all others.

The Galileo Project, along with NASA and the rest of the world's scientific community, must anticipate all such possibilities. There are compelling reasons to imagine interstellar space as home to more trash than traversing potential visitors. There is a compelling, if sobering, reason to imagine the opposite. Humans are already confronting mounting evidence that civilizations reliant on consumer-to-garbage manufacturing are not promising candidates to ascend very far up the ladder of civilizations.

The Galileo Project means to define the vocabulary, the frame of understanding, of our newly established interstellar civilization. It aims to do so even as governments are, deliberatively or not, framing their own understandings. That most of humanity is unaware of the fact that it has entered a new era, that the cultural norms of centuries of terrestrial habits still predominate, and that science and the scientific method remains the practice of a relative few of us is unsurprising. The foremost objective of the branches of the Project is to use science and its methods to accumulate data that allow us to move all unexplained phenomena that suggest UAP and ETC into the bins of the explained. It accepts the possibility that every such phenomena might eventually end up in bins for the naturally occurring or the human-engineered. However, it also anticipates that there will remain some phenomena that will resist fitting into either bin. For these, the Project anticipates building or aiding in the construction of instruments that allow us to only observe such phenomenon, or perhaps to image them, or perhaps to retrieve them for study. For that to work, we will need to embrace a new and unfamiliar set of expectations.

Lastly, we must make a virtue of our immaturity. As a civilization inhabiting a recently formed planet in a late arriving Solar system, we run the risk of always being too much like cosmic children for any curious ETC to find us worthy of their help.

With that in mind, to ease our ability to embrace new expectations and unfamiliar tasks, we should take prudent advantage of what is familiar. Which is why I've come around to the notion that, viewed through a glass scientific, the Galileo Project can perhaps be best understood as like the first crew of that iconic fictional effort to go where no human has gone before, the USS *Enterprise*.

3

NEW TELESCOPES FOR EXTRATERRESTRIALS

THE ROOFTOP OF THE Harvard College Observatory is about as unlike the bridge of the USS *Enterprise* as can be imagined. It has a stone patio feel, complete with a wooden picnic table for faculty and graduate students to use at lunchtime in nice weather. I am certain, however, that if humanity ever does construct a fleet of spacecraft, whether powered by technology we already have, lightsails, or perhaps some technology we reverse engineer from an ETC's artifact, such as gravitational propulsion, it will owe a small debt to the work done atop that roof.

It was July 2022 and we were hours from the Galileo Project's first, and humanity's first, UAP-specific observatory going live. The staff at work did not look much like the characters from science fiction fantasies. These were volunteers, mostly but not all undergraduates and graduate students. If they shared a uniform it was T-shirts and casual pants given the warm weather. And among the tools they had at hand were potato chips and an Allen wrench. Another was an umbrella, held aloft to shade workers and equipment.

We were nearly at a year postlaunch of the Galileo Project. Despite planning and hopes, only two of the seven UAP observatory's instruments were expected to go live that late afternoon. And to help make that happen, a half-dozen men and women were busy in the nearly 100-degree-Fahrenheit heat. One of them had arrived with snacks—the chips—and the wrench, which she put to immediate use in tightening a bolt in the inner workings of the Infrared and Near Infrared Camera Array. To do this, its protective dome had been removed. And to better allow her to do the necessary work, a black umbrella was held high to block out the Sun.

Exposed was a custom suite of seven FLIR Boson 640 cameras arranged radially to provide a full sky view of the surroundings, one Zenith ZWO all-sky visible/NIR wide-field camera, and one Zenith FLIR Boson 640 camera. Usually housed safely under a twenty-inch-wide fiberglass dome and cooled by eight sunshades and a filtered airflow fan, this afternoon it was open for tinkering and fine-tuning. Once closed and latched, for simplicity's sake this combination of observational instruments had been code-named the Dalek.

Near to it was a cylindrical Alcor Visible All-Sky camera, which ensures that the entire sky is continuously monitored. Both the Dalek and the Alcor camera sit atop a sturdy five-foot by three-foot metal work bench, with two bracing struts to a leg, all of it held securely to the roof by weights placed on either side of a lower cabinet that contains a computer responsible for running object detection and transmitting data. Built to be exposed to all elements for months or years at a time, the cabinet "is an electromagnetic interference (EMI)/radio frequency interference (RFI) shielded, ingress protection rated (IP65), weatherproofed enclosure."

I have watched relatively few *Star Trek* episodes, but I believe that quote would stand in very nicely for the sort of explanation the fictional engineer Lieutenant Commander Montgomery "Scotty" Scott would declaim whenever something was wrong with the inner workings of the *Enterprise*.

Our crew of interstellar explorers that particular day in July was a motley one. One volunteer, a lawyer by training but an astrophysicist by passion, was in charge of the umbrella. A graduate student from Wellesley College worked the Allen wrench. A crouching Harvard graduate student was adjusting settings within the cabinet, while two others looked on, and the spouse of one of the volunteers offered moral support from a shaded portion of the roof. Spread across the available surface of that roof was a fisheye camera atop a fast-slew mount and controlled by a targeting computer, an acoustic monitoring omni-directional system (AMOS), and a NPACKMAN, or a New Particle Counter K-index Magnetic Anomaly instrument designed specifically for the Galileo Project. Each instrument does specific work in gathering data about aerial phenomena. A camera for visual data, AMOS for acoustic data, and the NPACKMAN for data on environmental conditions and space weather near the observatory. Additionally, there was a dish for monitoring passive radar and, hidden a few floors down, a target coordination and data marshal system into which all the collected information is fed and analyzed. Taken together, they made up the Project's UAP observatory.

For months it stayed atop the Harvard College Observatory's roof for debugging and improving, after which it was deployed to rural Massachusetts. With every passing hour, the data collected and the underlying computer programming and analytical algorithms improved.

I think of the Dalek, and the women and men who conceived, designed, and built it, when I am asked, What will proof of an extraterrestrial technological civilization look like? This question was put to me repeatedly in the wake of 'Oumuamua's passage through the Solar system. My inability to be definitive that 'Oumuamua was or wasn't an ETC's artifact was a stumbling block. Distilled, my hypothesis was that among the objects that would behave as 'Oumuamua did we must include a non-terrestrial, artificially manufactured artifact. Whether or not it was a relic of an ETC was a matter of probabilistic conjecture made from frustratingly finite data. Some among the media, and even some of my fellow scientists, found in this uncertainty sufficient reason for rephrasing the question, Are we alone? in near accusatorial terms. *That's* your evidence for we're not alone? The spirit behind that accusation seemed to be, Give me a seat in a flying saucer and maybe I'll start to believe you.

Of course, the question, What will proof of an interstellar technological civilization look like? has an easy answer that is robustly substantiated by evidence. It will look like us. We are a civilization that has sent, and is sending, its technology out into interstellar space. We can take a sort of comfort from that fact. There is going to be some, however distant, common ground between human civilization and the first ETC we discover. This arises from the fact that when we discover another civilization's artifact, we can be confident that at some future point they, too, passed through a potato chip and Allen wrench phase.

The Dalek, the AMOS, and the NPACKMAN remind us that all scientific, technologically accomplished civilizations start with data. As your data improve, the questions left unanswered are fewer, and the technology needed to pursue the remaining mysteries become more specialized. With luck and with more specialized technology the civilization itself improves. The path to proof, I believe, is also the path to our better future selves, and it is a path followed through data gathering. Evidence of our progress is often measured by the winnowing of hypotheses and not eureka discoveries.

It's a bird, it's a plane, it's a UAP. Or is it a weather balloon, a thunder cloud, a high-altitude human-built craft for spying, a recreational drone? The UAP observatory I just described was built to learn how to distinguish among these. The purpose of that suite of cameras and observational equipment is to infer the kinematics of objects in the sky from their "tracks," or a time-series of localizations of an object in three-dimensional space. To accomplish this, you need at least two cameras separated by a known distance and calibrated to point in known directions. If an object is visible to both cameras, and if they are able to take multiple samples at the same time, then it becomes possible to reconstruct the image's 3D track. With those data, we can infer the velocities and accelerations of an object.

Once that information is fed into a suitably powerful AI computer programmed to account for a range of airborne objects and atmospheric phenomena under various weather and lighting conditions, we can begin to move UAP over to the EAP column.

What follows is the fun stuff of science. There is the problem to define, namely collecting data for the scientific study of UAP. There is the math to consider, namely the event-driven algorithm necessary to collect useful data from remote cameras for the purposes of object localization and identification. Within that is the need for a general triangulation error analysis to be used in gauging instrument performance "based on parameters such as baseline length, object distance, and object angles with respect to observers."

These considerations lead directly to hardware and its implementation. What technology is needed to gather the data that begin to usefully address the defined problem? In the case of the Galileo Project's first UAP observatory, almost all of it was off-the-shelf cameras and tools. Not off the shelf, however, was the material needed to house and protect the equipment from a wide range of wind, temperature, humidity, and illumination conditions. But what needed to be uniquely manufactured was for the most part the stuff of metal and wood shop construction.

The observatory's governing software also needed to be custom-made. Some of that was easy, such as the calibration of visual cameras. Open software that uses a chessboard pattern for calibration is readily available and works as well for the IR and NIR cameras. Global positioning systems found on smartphones can compute the translation vector of each camera, and we have calculated that they are accurate enough to perform within an acceptable level of localization error.

It was more challenging to create the artificial intelligence that we are training to identify objects. To help with that training, the Galileo Project developed AeroSynth, a tool that produces synthetic images similar to those the telescopes will capture. That synthetic data will be combined with real images to allow for the creation of larger, more complete data sets that will help ensure the observatories are capable of understanding the various kinds of conditions and phenomena found in Earth's atmosphere. The result will be an AI filter that will flag only those images that warrant further scientific investigation.

To complement the Galileo Project's UAP observatories, we are also proposing SkyWatch, or a passive multistatic radar network based on commercial broadcast FM radio transmitters. Such a network, spaced at geographical scale, would enable the estimation of the three-dimensional position and velocity of objects at altitudes up to eight kilometers and at velocities of around Mach six. Not only will SkyWatch be able to locate fast-moving objects, but locate them in all dimensions, an important consideration if one of these objects is able to turn 90 degrees on a dime and head off in a different direction. Once the object's positional information is gathered from the SkyWatch radar, those data will support the Project's wide-field optical cameras in target acquisition and monitoring. What SkyWatch brings to the effort is a detection range of 150 kilometers, which is an order of magnitude greater than that of the observatory's wide-field optical instruments. Relevant is the fact that the deploying of the most basic version of SkyWatch, a simple receiver solution, will be inexpensive. Much as you can drive a car from California to New York and never lose an FM station, similarly SkyWatch could extend its visualization of UAP.

Another ready-to-hand technology are satellite surveys of large areas of Earth. The Galileo Project is partnering with Planet Labs, which maintains more than two hundred Earth-imaging satellites, to photograph the entire landmass of the planet daily. The motion of an observed object captured in a single-frame image leaves a characteristic signature. Working from Planet Labs' SuperDove satellites, which image in seven spectral bands in the optical and one in the near infrared, the Project is exploring the feasibility of their use in identifying and characterizing UAP. Our research has demonstrated that "at least two characteristic signatures of motion in Earth observation images made with push frame scanning" provide a method for estimating motion and for estimating velocity.

The Project is now developing software that makes use of pattern-recognition techniques to automatically detect any moving objects in Planet Labs images.

Off-the-shelf cameras. Open-sourced software. FM radio transmitters of opportunity. A private company's growing stash of satellite images. None of it particularly expensive. Collectively, they will help the Galileo Project answer the Pentagon's call to aid in detecting UAP for threat identification. Identifying threats, of course, is not the Project's primary purpose or interest. Rather, as scientists active within a nascent interstellar civilization, we are scientifically curious about other ETC. And that curiosity is defined initially by the extent and quality of the data we gather, analyze, and share. From that vantage point, the Galileo Project is in the business of opportunity identification.

Most, and very likely all, such opportunities will be ones we initiate.

Here, again, there is limited but real value in returning to the fictional world of *Star Trek*. In its original series, dating back to the late 1960s, *Star Trek* introduced viewers to the Prime Directive. This was a generalized rule that prohibited ships and crew of Starfleet from interfering with alien civilizations. This was articulated as a matter of ethics. Altering a civilization's development, even if it appeared stuck or heading toward extinction, was deemed amoral. Almost immediately, the show's writers realized there was far more dramatic entertainment in instances where the directive was ignored or bent than in its assiduous adoption.

For more egotistical reasons, humanity has already ignored the directive. Both Voyager craft were sent out with the explicit hope of their encountering and influencing extraterrestrial civilizations, if only by providing proof that they are not alone. It is also probable that no such directive is needed. Highly advanced technological civilizations will be composed of intelligences that, but for a smattering of academics, will be uninterested in the struggles of immature civilizations. After all, it is generally true among humans that the more immature among us garner the least amount of interest from the more mature. This is why I think it doubly plausible that the manner and extent of our search for ETC near-Earth artifacts can prove successful. The first reason is the most obvious. As anyone who has misplaced their car keys knows, the more concerted the effort to find them, the more likely, and quickly, you will. Another reason is an ETC may only wish to be discovered by a civilization capable of doing so.

Research into discovering the nature and intent of extraterrestrially manufactured UAP presupposes an extraterrestrial civilization sufficiently curious to surveil us but insufficiently impressed to say hello. It is not so much the implausibility of a civilization so coy or ambivalent toward humanity that gives me pause as it is the technical difficulties that must be surmounted. Sending a small craft near a potentially inhabited planet to gather data to be returned to a home planet isn't the stuff of science fiction, just science. As a matter of technology, humans have the know-how to send a quite small craft (about the size of an iPhone) to take pictures of Alpha Centauri B, a planet in the habitable zone of the red dwarf star Proxima Centauri. What is more, it is within our ability to have that craft reach Alpha Centauri B in about two decades and return images back to us in about another five years—or well within a human lifespan. What that craft would *not* do, however, is enter Alpha Centauri B's atmosphere, let alone zigzag along its horizon, or hover in one place despite headwinds. Such a craft vastly exceeds our technological abilities and defies a few laws of physics as we know them. Our far humbler craft would instead pass between planet and star, gathering what images and information it could before its trajectory and near light-speed velocity took it past the range of its instrumentation. If an interstellar civilization of our immaturity can do this, an interstellar civilization even just a few grades more mature could send more sophisticated and simply more (quite small) craft to our Solar system. Such craft would still be unlikely to enter our atmosphere, or survive the encounter if they did. Which is why the Galileo Project's plan to encourage an intercept mission with a sufficiently intriguing interstellar object before any such encounter is among its most hopeful.

I describe the effort as dating the next 'Oumuamua. I use the analogy in part because it helps take something our civilization treats as exceedingly unlikely—the discovery that we're not alone and possibly even deserving of extraterrestrial attention—and frames it as something most of us treat as challenging but common—the discovery that another human is interested in spending up to a lifetime with us. In both cases, discovery begins by gathering sufficient information about objects of interest.

We have compelling reasons to believe that our galaxy contains a large population of planetesimals, or material ejected into interstellar space as a result of planet formation and migration. That some of these will pass through the Solar system becoming interstellar comets, like 2I/Borisov discovered in 2019, is likewise expected. Most such comets will likely be uninterestingly similar to all the locally generated comets we have cataloged. A few such comets, however, will exhibit decidedly outlier properties. One example is, of course, 'Oumuamua, whose properties were outlier enough to allow it to be plausibly considered as a manufactured artifact. But we can only intercept what we know, with sufficient warning, is out there, and for the past decades that was next to zero. With the forthcoming Vera C. Rubin Observatory's Legacy Survey of Space and Time, however, we can estimate with high confidence that we will soon be identifying one to ten interstellar objects of 'Oumuamua's size and demonstrating some or all of 'Oumuamua's properties every year. The Galileo Project's ISO branch will consider the LSST pipeline of data on interstellar objects like a dating app: routinely, we will swipe to the left until we find the object that is sufficiently anomalous for us to pursue it.

A later cohort of Galileo Project staff will have to determine whether an interstellar object is sufficiently interesting to warrant intercepting it. They will need instruments that allow for high confidence inferences about its composition, reflectivity, and surface roughness. They will study its emission spectroscopy with care, for along with reflectance it will provide among the most valuable insights into what may be entering the Solar system. Planetary space probes typically use spectrometers to provide spectral maps of surfaces, information about the surface minerals, extent of space weathering or irradiation by high-energy particles, and the extent of surface ice. The same instrumentation will help in identifying potentially artificial materials. Based on the human interstellar technology we are familiar with, we would first seek markers of solar cells, circuit boards, and aluminum. NASA's Optical Measurement Center using an Analytical Spectral Devices field spectrometer was additionally able to note absorption features that were likely the result of water and indications of the presence of organic content. Another possible discriminator between a natural object and an artificial one is its albedo, or the fraction of the total solar radiation reflected by the object back to space.

How many 'Oumuamua-like objects we can identify, how quickly we can identify them, and how prepared we are with instruments and craft to survey and image them is the now theoretical but eventually practical work of the Interstellar Object branch of the Galileo Project. Soon, it will receive a strong nudge from the completion of the Vera C. Rubin Observatory to shift from the theoretical to the practical.

Though not built to pursue questions about the existence of extraterrestrial, technologically advanced civilizations, the Vera C. Rubin Observatory is a tool like none previously available to humanity to aid in our observation of near-Earth objects. Using a telescope equipped with an 8.4-meter mirror, a 3,200-megapixel camera, an automated data processing system, and an online public engagement platform, the Rubin Observatory is designed to conduct a decade-long optical survey that will provide a global population of scientists an unprecedented look at "every part of the visible sky." While mapping the Milky Way, probing the nature of dark matter and dark energy, and cataloging the Solar system, the Rubin's 500-petabyte images will also vastly increase our discovery of interstellar objects. And recent NASA missions have also demonstrated our competence at landing atop and mining passing space rocks.

Everything human civilization needs to rendezvous with another 'Oumuamua-like object is in our toolkit. The question is, Will we use these technologies for this purpose?

INTERSTELLAR SPEED DATING

To meet the next 'Oumuamua, the first thing we'd need is a really fast craft.

Catching up with such an object would require an interceptor space rocket capable of between ten and five kilometers per second impulsive thrust. This would allow the interceptor to approach the next anomalous interstellar object with properties similar to 'Oumuamua to within a distance of one thousand kilometers. Armed with a telescope measuring half a meter in diameter, such a spacecraft would allow us to distinguish meter-sized features over the twenty seconds that the encounter lasts. The more instruments we add, the more we can learn.

There are trade-offs. The more instrumentation, the heavier our rocket becomes, and the more fuel required to have it reach requisite speeds. And then there is the expense of the instruments themselves, along with, of course, the rocket and its fuel. If this is beginning to sound prohibitively expensive, you are right. While it is possible to send the astronomical equivalent of a sports car into space, as Elon Musk did, it is an unjustifiable expense; sometimes it really is the better part of wisdom to not order the most expensive item on the menu.

Given costs and efficiencies, here's what the Galileo Project's Interstellar Object team is currently defining as the kind of equipment necessary. In addition to the main telescope—a relatively inexpensive object that is common enough to be considered as off-the-shelf available—the interceptor would carry a wide-field camera to guide its trajectory. And it would also include a visible-near infrared multi-spectral imager like the one named "Ralph" that NASA included on the New Horizons spacecraft it sent to Pluto. Our Ralph would allow us to study the composition of the object's surface. We can anticipate that the effects of space weathering will have left that surface pitted, perhaps battered. But our sensors would be sufficient to distinguish an artificial object from a natural one.

Perhaps the most expensive decision will be where we place our interceptor. Location matters. A rocket sent from Earth to meet up with an interstellar object represents many degrees greater difficulty of timing and calibration than one sent from a launchpad at L2, or the second Lagrange Point (LP), which sits one million miles from Earth. There are five such points and their determining feature is a balancing of gravity such that a spacecraft or telescope placed at any LP will remain in a fixed position relative to the Sun and Earth with minimal need of course correction.

Even when launched from L2, we will need to be patient. Had such a rocket been launched on July 25, 2017, for instance, it would have taken 83.38 days before it crossed paths with 'Oumuamua on October 25, 2017. Still, the reward for our patience would have been immense. By October 26, we would have had highly conclusive data as to whether 'Oumuamua was a hydrogen or nitrogen iceberg, an ultra-porous aggregate a hundred times less dense than air, a planet fragment that was tidally disrupted, or a piece of extraterrestrial technological equipment. It is often said that a picture is worth a thousand words; in fact, a high-resolution image of 'Oumuamua would be worth more than the billions of words spoken about it.

Fortune would play a role in how much data we collected. If we were lucky—and luck is always a part of any data—in addition to our twenty-second view of the object's meter-sized features, the interceptor would also have about three minutes of observations with a ten-meter resolution. It would be the ultimate in speed dating, especially given the likely price tag of approximately $1 billion.

It seems churlish to put a price tag on a momentous scientific study of an extremely rare object. After all, by the time we have limited our interest to 'Oumuamua-like objects, we have already focused our scientific instruments on the most outlier phenomena among interstellar travelers. We will either discover a never-before-seen natural object, which will present empirical data for astrophysicists to ponder as they seek to explain the making and unmaking of the matter that makes up the Universe. Or, we will discover a never-before-seen artificial object. What is certain is that as of the time I write this paragraph, human civilization has nothing in its fleet of spacecraft built to find out which.

Finding evidence of another civilization in the Universe starts with our willingness to begin searching—just as finding a friend or partner rests on an openness to begin looking. The likelihood of success turns a great deal on the extent and nature of the effort expended searching. A civilization awaiting a partner on a white horse to show up at its front door without prompting or effort is a civilization living out a fantasy akin to *Sleeping Beauty*. And while numerous myths and fairy tales allow for kismet and magic to match the deserving lovelorn, this never happens in the domains of science. Science leaves zero room for the explanation "and then a miracle happens here." There are always visible gaps in our theory and our data. While culturally casting your search for a lifetime partner in terms of theory and data is not recommended, advancing science in any other way is, in a word, unscientific. For a scientist to behave otherwise would constitute the equivalent of Enrico Fermi insisting extraordinary evidence present itself just because he wished it would.

Of course, no scientist behaves this way, and when he was helping to construct the first nuclear reactor, least of all Enrico Fermi. Hopeful expectations didn't gift the United States with two nuclear bombs at its disposal to end World War II. Years of work and vast expense did. Similarly, wishful thinking is the weakest tool currently being used in our search for the nature of dark matter. Most of us know that while Mr. or Mrs. Right might appear on your first-ever date in life, the odds are greater that some percent of Wrongs will need to be encountered first.

This is why I believe Carl Sagan's declaration that, when it comes to the existence of extraterrestrial civilizations, extraordinary claims require extraordinary evidence is poor advice, not just for scientists but also for the lovelorn and hopeful. Sagan's wife, the very-accomplished scientist Ann Druyan, beautifully reminisced following his death in 1996, that "Every single moment that we were alive and we were together was miraculous—not miraculous in the sense of inexplicable or supernatural. We knew we were beneficiaries of chance." This can sound like a recasting of Sagan's admonishment, stipulating that extraordinary evidence—true love—requires extraordinary luck. Chance, pure chance, in the sense of rolling two dice thirty times and getting snake eyes every time, is indeed extraordinary, but not as a matter of advancing evidentiary claims. As anyone seeking a partner knows, there is luck involved in finding the right person but there is also a deliberative practice in putting yourself out there and getting experience and knowledge, and trying and trying again and again.

My guess is most couples, whether first-time lovers or lifetime partners, encounter far more ordinary claims backed up by ordinary evidence to encourage each to consider meeting up again sometime soon. And not infrequently, carefully accumulated evidence leads to the failure of a relationship. Disproving the hypothesis that one person is the right match is often necessary to support the extraordinary discovery of one who is far better—true love in fact. After all, Sagan was twice divorced when he and Druyan got married.

Therein lies a universal nugget of insight. Similar to knowing what scientific hypothesis you wish to advance, knowing who you want to date is always a function of ignorance; first you need to know the pool of potential candidates. As humanity now begins the work to seek a potential partner civilization, it is a given that humanity is woefully ignorant of the Universe's possibilities.

After all, it is only within the last thirty years that we discovered planets outside the Solar system. The Universe, it turns out, is teeming with super-Earths and mini-Neptunes, the most common types of planets in the Galaxy. Along with their discovery came frustrations: they all sit at enormous distances from Earth and our own Solar system. The prospect of visiting one in person is remote, both generations off in our technological abilities and quite likely requiring further generations of time travel. Pining to visit in a lifetime one of these exoplanets is similar to pining for the actor on the movie screen for a lifetime. Based on what we currently understand about physics and rocketry, certain interstellar relationships are destined to be confined to what we encounter in images.

But this doesn't make them unimportant or uninfluential, just noninteractive.

The Breakthrough Starshot Initiative, founded by Yuri Milner, Stephen Hawking, and Mark Zuckerberg, and of which I have the honor to chair its advisory board, predates the Galileo Project by five years. It is a research and engineering project that has found a means of sending a craft, perhaps a fleet of them, to other star systems. The closest star, Proxima Centauri, is 4.24 light-years away from Earth. Put in context, the distance light travels in 4.24 years would take our current fastest spacecraft 100,000 years to cover. This is why the Breakthrough Starshot Initiative intends to design a lightsail-powered nanocraft that, when propelled by a sufficiently powerful beam of laser light, will reach speeds of 130 million miles per hour. At that speed, such a craft—what has been nicknamed a StarChip—could reach the Centauri system at a fifth of the speed of light, image the exoplanet Centauri B, and transmit the data back to Earth in merely a quarter of a century.

Humanity's first interstellar fleet of spacecraft will look nothing like the ships from *Star Trek*. Each StarChip will be tiny, weighing approximately a few grams. Their instrumentation will depend on lightweight microchips powering small but affordable cameras capable of resolutions of at least 200 by 200 pixels. The craft's lightsail will need to be similarly lightweight but also highly resilient. After all, the plan is to direct 100 gigawatts of laser light at it. Only a few materials promise to not be instantly vaporized, among them graphene. Let go of visions of huge craft populated by hundreds, perhaps thousands of crew members. Imagine instead a very sophisticated iPhone attached to a small graphene lightsail passing a planet at a significant fraction of the speed of light and you have a fairly accurate image of what is proposed, and attainable. For the Galileo Project's interest in UAP and ISO, the Breakthrough Starshot Initiative has immediately relevant insights by way of transitive logic. That humanity could now send one, and perhaps thousands, of StarChips out into space tells us that any other ETC will very likely have passed through its own phase of launching nanocraft.

INTERSTELLAR ARCHAEOLOGY

It is no accident that just as humanity approaches the technological prowess to send out interstellar probes, we begin to consider what it will take to seek both near-Earth objects and near-Earth probes. To date, it is a truth of our civilization that our curiosity is usually bound by our appreciation for the leading technological edge of the current generation.

For decades, the Search for Extraterrestrial Intelligence (SETI) was focused on radio or laser signals sent from distant civilizations. Those decades correspond to when humanity was reaching maturity with its radio technology and was beginning to establish its mastery of laser signals. SETI's traditional approach, however, remains the equivalent of waiting for your phone to ring. To receive an electromagnetic signal, we need the sender to transmit it exactly a light-travel-time ago with similar communication technologies to those we developed over the past century. The odds of this happening are mind-bogglingly long. The odds are less than one part in a hundred million—the fraction of time spanned by radio communication on Earth relative to the age of most stars. Most such signals would have escaped the Milky Way galaxy long ago and by now are faint undetectable glows billions of light-years away.

SETI's hoping for the phone to ring reveals our human civilization's youth to be a curse. The Sun formed in only the last third of cosmic history, which has spanned a total of 13.8 billion years since the Big Bang. Direct observations show that the peak of the star formation history in the Universe was 10 billion years ago. Hence, most stars formed billions of years before the Sun. Additionally, we know enough to estimate that within one, at most three billion years the Sun will increase its luminosity and boil off all oceans on Earth and sterilize its surface, most likely ending all life-as-we-know-it. Assuming similar circumstances throughout the Universe, this means that most Sun-like stars sterilized long ago their previously habitable Earth-like planets. If these planets hosted technological civilizations, then these cultures died or, more hopefully, migrated elsewhere. No matter. Human civilization wasn't around to hear their cries for help from billions of years ago.

Correspondingly, it is pointless to search for radio signals from these sterilized exo-Earths now. We arrived late to their party. It is over.

But there is also good news. There is an alternative search method that makes a virtue of our civilization's youth. It involves checking for physical packages in our mailbox. Even if radio-transmitting civilizations are now long dead, their packages might have piled up in our cosmic neighborhood, awaiting our sufficient interest.

If previously existing civilizations launched chemical rockets into interstellar space, these probes are still around us since they are gravitationally bound to the Milky Way galaxy. The escape speed from the Milky Way is 500 kilometers per second in the vicinity of the Sun. This is an order of magnitude faster than the limiting speed of chemical propulsion, and manifested by all the interstellar probes that humanity has launched so far.

These fortunate circumstances allow us to potentially find evidence for past civilizations, whose clock started ticking billions of years before ours. Our near future could well be influenced by their long-ago past.

This realization calls for a new research frontier of "interstellar archaeology," in the spirit of searching our backyard of the Solar system for objects that came from the cosmic street surrounding it. The traditional field of archaeology on Earth finds relics left behind of cultures that are not around anymore. We can do the same in space.

Our search for interstellar artifacts could uncover either functional devices or defunct space trash. Functional devices deliberately sent in search of life, as would be the case for our StarChips, would most likely focus their trajectories toward the habitable regions around stars. As a result, they could be far more abundant in the vicinity of Earth than on average in interstellar space. Given that the extent of the Solar system is a hundred thousand times bigger than the Earth-Sun separation, the local density enhancement of such life-seeking devices could be some fifteen orders of magnitude. Space trash, on the other hand, is likely to be as broadly and carelessly discarded as it is on a thoughtless species' home planet.

There are two primary ways to find interstellar objects. We can look for them under the "lamppost" of the Sun, the bright source of light that illuminates the darkness around us. Just as it is easier to find a dropped object under a lamppost than it is in a dark alley, so it is easier to see an object's luminosity as it passes near the Sun. This is how Solar system asteroids or comets are routinely found. How quickly they pass is also a tell. Interstellar objects would be faster than Solar system objects because they are unbound to the Sun.

The second way to find interstellar objects is using the Earth as a fishing net. We can search for objects that collide with it at high speeds. Spying these will be easier. They will produce their own light as they burn up as a result of their friction with air.

At the birth of SETI in 1960, the visionary radio astronomer Frank Drake presented his justly famed eponymous equation. It presented a way to estimate the number of extraterrestrial civilizations with which communication might be possible, or the number that could be communicative via detectable signals. Interstellar archaeology can rest on different assumptions. With the search for relics of other technological civilizations the continued existence of the civilization isn't at issue. Rather, it is the possible existence of its artifacts. In a real and very hopeful way, the search for extraterrestrial artifacts is much simpler.

Consider two equations. First, the Drake Equation used to estimate the number of extraterrestrial societies detectable by their electromagnetic emissions:

$$N = R_* \times f_p \times n_e \times f_l \times f_i \times f_c \times L$$

For our purposes, it doesn't matter what all the factors try to capture. Many represent best-guessed and most-hoped-for assumptions. They include the knowable (such as the average rate of star formation in our galaxy) and the unknowable (how long a civilization that is able to emit signals manages to survive). Humans have no instruments by which to gather any useful amount of data that would help us answer the latter. And this is why the Drake Equation, with its seven variables, of which a few could easily be close to or in fact zero, remains an expression of hope more than a guide to scientific work.

By comparison, here's an equation that could guide the Galileo Project's branch of research into whether or not we could intercept an extraterrestrial artifact:

$$N = n \times V$$

The number of artifacts, N, will depend on the number of extraterrestrial artifacts per unit volume, n, that exist within the population of objects sufficiently intriguing to merit our attention within a survey volume of space, V. Because of 'Oumuamua, we know no variable is zero. We are already aware that sufficiently intriguing objects exist in our Solar system. The work ahead is learning if any such object is an alien artifact.

Precisely because the second equation is a guide to a scientifically answerable question, we can spy the flaw in it. Everything will depend on what tools we design, build, and use to gather the data relevant to n and V. If we are insufficiently curious, if we refuse to design or build or use appropriate tools of discovery, then of course the zero value of N—no artifacts found—will remain. With this in mind, I offer a variation: $N = n \times V \times (1-O)$.

The variable O, with a value of 1, is intended to honor the ostrich's rumored preference to stick its head in the sand rather than confront unwanted, uncomfortable realities. In my experience, humans tend to overlook the value of O, just as we tend to forget that we are far more likely to ignore disquieting realities than ostriches are. By including the variable 1 minus O, I am allowing that the number of near-Earth extraterrestrial artifacts that humans ever discover will owe a direct debt to how hard some humans work to find them.

In the thought experiment I introduced in the first chapter of the alien civilization that located and caught Voyager 1, I believe the most significant difference between humans and my imagined extraterrestrials is the latter's default refusal to act the scientific ostrich. The speed with which human civilization will make progress toward answers to, Are we alone?, and, Is ours the only technological civilization in the Universe?, will be functions of our having different expectations of science, technology, exploration, of, in brief, humanity. After all, by virtue of birds, proof of flight was daily right in front of us for thousands of years before humans built a vehicle capable of allowing one of us to stay aloft. For nearly four hundred years, from 1485, when Leonardo da Vinci sketched (but never built) his Ornithopter, to 1891, when Samuel Langley's steam-powered Aerodrome flew for three-quarters of a mile, a few humans struggled, defied critics, and most often failed. And then, eureka! In under two years the Wright brothers went from a twelve-second success in 1903 to, by 1905, their Flyer III staying aloft for thirty-nine minutes and covering twenty-four miles. The question, What would hard proof of human flight look like? had leapfrogged to, What is required to keep humans aloft longer?

Our historians tend to tell the history of human flight as one of givens. Da Vinci had sufficient imagination, but he predated the political, cultural, and industrial revolutions the Wright brothers needed to own a bicycle shop of materials and the ability to tinker with them. Humans tend to presume our historical trajectory is not optimal but largely unavoidable. Because it took humans four hundred or so years to get from a sketch to a Flyer, then that must be about the time it takes for other civilizations to do the same. If true, it gives us all the more pressing a reason to seek, and hope to find, an extraterrestrial artifact. Either there are no homegrown shortcuts that a civilization can engineer to leap rungs of the ladder of civilizations at a time, or humans are disadvantaged in their civilization being unable to do so. Either suggests that the discovery of just even one defunct, partial object of extraterrestrial manufacture has the potential to do the equivalent of collapsing the centuries from da Vinci's sketches to your last transcontinental flight. That is among the hopes behind the expedition to locate IM1 off the coast of Papua New Guinea.

4

THE MESSENGER

EVERY NOW AND THEN, something really big strikes Earth and changes everything.

Earlier, I made passing mention to the Chicxulub asteroid, which is credited with bringing the age of the dinosaurs to an abrupt end 66 million years ago. Most estimates put it at "about the size of a mountain." That reflects guesswork based on its impact crater, most of which is offshore of Mexico's Yucatan Peninsula and is approximately 110 miles in diameter and 12 miles deep. The energy released would have vaporized rocks, triggered massive tsunamis, sparked global firestorms, and spread dust and sulfuric soot throughout the atmosphere.

For the most part, twenty-first-century humans took the lesson: what the dinosaurs didn't know, killed them. In 2005, the United States Congress tasked NASA with tracking 90% of all near-Earth objects 140 meters or larger. It was to be done by 2020, though by that year only a third of such objects had been identified. Very soon, scientists will attempt to get closer to fulfilling the 2005 directive, courtesy of the Vera C. Rubin Observatory's Legacy Survey of Space and Time. The same instrumentation the Galileo Project proposes to use to identify the next 'Oumuamua-like object will also help humanity identify concerning meteors.

The final branch of the Galileo Project, however, is interested in smaller meteors, or, to be more precise, interstellar meteors, starting with IM1.

Small objects enter Earth's atmosphere all the time. NASA guesses that about 100 tons of material, mostly dust and gravel, do so daily. Few of them are interstellar. And few of them are of a size to make an impressive explosion. On December 18, 2018, for example, a meteor spotted off the Kamchatka Peninsula on the far eastern edge of Russia produced a blast 16 miles above the Earth's surface that released ten times the energy of the Hiroshima atomic bomb. It was estimated to be about 10 meters in diameter and weigh about 1,600 tons, and to be the third largest meteor to impact Earth since 1900.

Of course, only a few objects per decade that strike the Earth's atmosphere release energies that can be compared to atomic bombs. Most meteors leave little to nothing for subsequent discovery. The ones that don't disintegrate are usually made of extremely tough materials. Unsurprisingly, the largest meteors ever to be discovered are almost entirely made up of iron, among the strongest naturally occurring metals. Humans long ago learned how to manufacture stronger alloy metals, such as steel. Nature hasn't the intentionality of a metallurgist and goes about creating strong metals more haphazardly. The Willamette Meteorite, weighing in at 15.5 tons and on display at the American Museum of Natural History in New York City, is metallic iron likely formed from the collision of two protoplanets orbiting the Sun billions of years ago. It is surmised that the Willamette is what was left of one planet's iron-nickel core after its encounter with Earth. Just the fact that it is iron, however, makes it rare. Iron meteorites make up about a twentieth of all space rocks arriving on Earth.

As my comments about the Chicxulub, Kamchatka, and Willamette meteors indicate, scientists have become ever more skilled at sleuthing out the origin and life stories of the rocks that hit Earth, a fraction of which leave discoverable remnants.

A word about the word *meteor.* It defines both a phenomenon—the streak of light caused by a space object heating to incandescence from friction with the Earth's atmosphere—and, more colloquially, the object itself. The latter usage, however, isn't precise. There are technical terms that scientists use to describe space rocks that may collide with a planet. *Meteoroid* is the term for a space rock before it encounters any atmosphere; *meteors*, the heated and disintegrating rock encountering the atmosphere; and *meteorite*, what remains of the rock on hitting Earth. Space is full of meteoroids, only a fraction of which become meteors, and only a fraction of those leave discoverable residue in the form of meteorites. On January 8, 2014, IM1, the first identified interstellar meteor, did precisely that, with its fragments (the largest of which is about half a meter in size) coming to rest 1.7 kilometers below the surface of the Pacific Ocean in the waters off Papua New Guinea.

When it encountered the Earth's atmosphere, IM1 produced a bright fireball and released energy equivalent to a few percent of the atomic bomb dropped on Hiroshima. This was enough to be detected by satellites and ground-based sensors. From data gathered by the United States Department of Defense, we also know that IM1 had an unusual light curve, which produced three separate flashes that were separated from each other by about a tenth of a second. The measured motion for IM1 can be used to calculate the altitude of the three flashes and the ambient density of air where they occurred.

When a meteor traveling at supersonic speed moves through the air, its frontal surface area is subject to friction. The force per unit area equals the ambient mass-density of air times the square of the object's speed. The resulting ram pressure slows the object, which will disintegrate if the pressure exceeds the yield strength of the material it is made of.

What follows is technical, but important. It is the science behind our confidence that IM1 is an outlier, something rarely encountered and deserving our concerted effort to retrieve and study it. Succinctly, where and how IM1 broke apart tells us it was composed of materials stronger than 95% of known meteors. With the conservative estimate that IM1 traveled at 44.8 kilometers per second, we place the three explosive flares, or moments when its disintegration was powerful enough for our instruments to note it, at altitudes of 23, 21, and 18.7 kilometers. We can estimate the atmospheric density of each flare site and so arrive at the ram pressures corresponding to the three explosive flares. Measured in megapascal pressure units, these are 113 MPa, 145 MPa, and 194 MPa, with ram pressure increasing as the meteor's speed decreased.

Because humans have cataloged a history of meteors, we have an established range of yield strengths for carbonaceous, stony, and iron meteorites. The upper end of yield strengths, reserved for the strongest materials, corresponds to iron meteorites and is about 50 MPa. And that means IM1's minimum yield strength of 113 MPa exceeds that of iron meteorites by a factor of 2. An improved acoustic localization of IM1 implied a material strength larger by another factor of 2-4.

The conclusion is that IM1's composition makes it tougher by at least an order of magnitude than all other 272 space rocks which slammed into Earth over the past decade in NASA's fireball catalog, CNEOS. Every now and then, something very small hits Earth and changes everything, and that just might include IM1.

SCIENCE AND SECRECY

In several respects, IM1 has already changed many things for the better. For one, it heralds humanity's new, recently begun future. If 'Oumuamua was a scout, IM1 is a messenger. And just as with 'Oumuamua, how much changes and how much for the better depends on how humanity responds.

The 2017 discovery of 'Oumuamua, again, provides a useful start point for humanity's interstellar future. With the debate its passage initiated, we began to think differently. I, for one, found myself wondering if 'Oumuamua was indeed the first interstellar object ever detected by humans. I knew that NASA's Center for Near Earth Object Studies had compiled a catalog of them with the Pentagon's help. Once it had become clear that NASA would not meet its congressional asteroid-tracking mandate, it struck a deal with the United States Space Command, or that branch of the armed services that oversees military operations in outer space. NASA knew that Space Command had access to data, collected by Pentagon-owned resources, that would go a long way toward helping it complete its catalog of near-Earth objects. Because some of the United States military's most sophisticated satellites were used in compiling the catalog, only some of the data NASA relied on was made publicly available.

More data were not released because decision makers higher up in the governmental bureaucracy feared such information would help adversaries learn just how sophisticated our military satellites are. What was available for public review, however, included the objects' velocity and position at the time of impact with Earth's atmosphere. Those data alone could be sufficient to determine if a meteor is of interstellar origin.

Curious, in 2019 I asked my then undergraduate student Amir Siraj, who had already been working with me on 'Oumuamua, to review CNEOS's catalog and to calculate the past trajectory of the fastest meteors in it, starting from their detected position and velocity at impact, and taking account of the gravity of the Earth, the Sun, and all distant planets within the Solar system. We were hunting predecessors of 'Oumuamua.

We set the parameters for speed and trajectory of what we would find interesting well beyond the norm, wanting to winnow the catalog down to objects most probably of interstellar origins. Within short order, Amir reported his findings. The trajectory of the fastest object detected put it on course for a head-on collision with the moving Earth. To humanity, and NASA, such an object is the stuff of nightmares (and many Hollywood plots) and why governments are building planetary defense technology. But to Amir and me, who were hunting interstellar objects, it was of less interest. While its speed relative to Earth was notable, its origins were not. Our instruments gather sufficient data to allow us to model a meteor's path, which in turn allows us to state with high confidence if it originated within or outside of our Solar system. The fastest object

Amir identified was on a path that meant it was likely gravitationally bound to our Solar system.

The data concerning the second-fastest object, however, strongly suggested that it was unbound to the Sun. In other words, its path suggested that it originated outside of our Solar system, that its origins were interstellar. The third fastest was possibly bound, but with uncertainties. The second fastest, now labeled IM1, riveted our attention. Amir's email to me summed up both the excitement and the finding: "We might have discovered the first meteor which originated outside the solar system!" So commenced a series of instructive frustrations.

First, there is the confusion of chronology. There are the dates on which data about asteroids and meteors are gathered. And then there are the dates on which humans make interpretive sense of the data, leading to "discoveries." Data about IM1 were gathered by our instruments on January 8, 2014, a small part of which was then made public in the CNEOS catalog. There those data stayed, uninterpreted by humans interested in interstellar objects. Data about 'Oumuamua were gathered by our instruments on September 9, 2017; forty days would pass before review of those data led to the discovery that 'Oumuamua was interstellar. With IM1, the data were collected in 2014, but humans—Amir and I—only discovered it to be interstellar in 2019.

Except it took about two years for our 2019 discovery to be declared a discovery in fact.

A few days after our discovery, we posted a paper summarizing our data and conclusion on the arXiv, an open-access archive of preprints, not-yet-peer-reviewed science, and simultaneously submitted it for publication to *The Astrophysical Journal Letters* . . . whose editors rejected it. They allowed that what we had found was highly suggestive, but wanted finer data proving that IM1 was interstellar.

While I thought what we had inferred from the publicly available data was suggestive enough, I fully understand that more, and more precise, data would strengthen the case. The irony is that everyone knew that more precise data existed, but that they were hidden behind walls of government secrecy. The very terrestrial concern that adversaries of the United States might discover the capabilities of our satellites trumped the scientific interest in whether an interstellar object had landed on Earth in 2014.

In an attempt to get to the truth, I wrote to a scientist who had sufficient security clearance to access the data and whom I met in my capacity as chair of the Board on Physics and Astronomy of the US National Academies, asked him to check our conclusions against it, and let me know what he discovered. He wrote back confirming that, yes, we had indeed identified the first interstellar object discovered by humanity. Elated, we resubmitted the paper . . . and were again rejected.

Why? The paper's editors concluded that the word of a scientist with sufficient security clearance to have reviewed the relevant data was not trustworthy. Again, I was sympathetic with their argument, and frustrated. Transparency of data is a touchstone to the scientific method. Trust, however, is a touchstone to human civilization. No matter, without transparent confirmation of the relevant data, Amir's and my finding was relegated to the dustbin.

And there the matter rested until something perhaps even rarer than an interstellar meteor occurred: the military decided to intervene to advance public science. A collection of scientists, members of the United States Space Force (USSF), NASA, and the White House's Office of Science and Technology Policy convened to discuss overlapping concerns, which included NASA's directive to identify near-Earth objects. Though meteors that had already struck the Earth were of only distant interest, Amir's and my paper, "Discovery of a Meteor of Interstellar Origin," was mentioned in passing. It was observed that our discovery was held in scholarly limbo because of insufficient data, and that the government in fact held the necessary data. This set off a chain of events that caused Joel Mozer, the Chief Scientist of Space Operations Command, USSF, to review data held exclusively by the Department of Defense. To their credit, this collection of scientists, generals, and policymakers recognized the worldwide importance of this finding and prevailed upon Lieutenant General John E. Shaw, USSF, to release a formal letter to NASA to confirm IM1's interstellar origins to a 99.999% confidence. The Department of Defense came to our defense, a stunning act from an organization far more conservative than blue-sky academia.

A SEISMIC CHANGE

You can hear interstellar space knocking on Earth's door if you type into your Web browser https://lweb.cfa.harvard.edu/~loeb/PNG.html. There you can listen to an audio rendering of the seismometer on Manus Island, Papua New Guinea, registering the impact of IM1. What you hear when you listen to it gives a hint at your level of hopeful curiosity at this dawn of our interstellar future. I hear an awesome announcement that a piece of the ancient Universe, whether of natural or manufactured origins, has arrived not at our doorstep but on our terrestrial lap. Others, however, might just hear two mundane, fleeting, flat notes: phhhphtht; phhhphtht.

If you are expecting to be entertained, you will probably be disappointed. But reality does not set out to entertain us. Nor does it stubbornly resist our efforts to fully understand its laws and properties. Reality isn't awesome, confounding, or deliberative. From periodic table to quantum entanglement, reality simply is. Awe is what humans enjoy, and sometimes suffer, during our slow, plodding appreciation of reality's laws and properties. For centuries, some of us have allowed our awe to direct us to undertake the science, exploration, and manufacture of the necessary technology to master more of the reality that surrounds us.

Among humanity's arsenal of instruments is the seismometer. Built to register and record ground movement, it is a simple piece of technology. Part of the device is fixed to the ground, part of it is suspended or made sensitive to the ground's movement. When, due to earthquake, volcano, or explosion, the ground moves, the suspended, sensitive part registers that movement. By written description, such a device was supposedly built in China as early as

130 AD. Ground movement resulted in one of eight dragons, positioned at different points around a metal vase, dropping a ball into a bronze toad's mouth, thereby pointing in the direction of where the earthquake occurred. I can imagine the awe felt by its inventor the first time a ball struck a toad.

Today, thousands of seismometers are positioned around the globe. For example, the United States Geological Survey (USGS) maintains numerous seismometers as a hedge against earthquakes, primarily. If you're so inclined, you can enjoy a spike of awe by visiting their website's real-time seismogram displays. There, at a glance, is the rustling of the Earth's rocky plates that humanity depends on as they incrementally move atop a layer of partially molten rock.

Fortunately, Earth's plates move slowly. Depending on the density of the rock, friction, and convention currents in the mantle, they travel between 5 and 75 centimeters a year. We only notice when plates collide along fault lines, such as the San Andreas Fault of California. But plate tectonics are not just of interest to Californians trying to decide where to live or what insurance to buy. On occasion, astrophysicists pay attention to it, too. I, for one, was mindful that the region around Papua New Guinea sits within the converging Australian and Pacific plate boundary zone. It is considered to be "one of the most tectonically complex regions of the world."

When NASA finally confirmed that IM1 was in fact the first interstellar object, Amir and I were briefly thrown into the media's glare, but I was occupied by a single, overriding thought. It wasn't just that we had discovered in the CNEOS catalog the data trail that led to identifying the first interstellar meteor, and that also informed our understanding of its rare material strength, but that we now knew where the remnants of that meteor lay. The Department of Defense located IM1's airburst at 1.3S, 147.6E degrees in latitude and longitude at an altitude of 18.7 kilometers. So, because of satellites and seismographs, we know the debris from IM1 now rests within a 10 kilometer by 10 kilometer grid of ocean floor near Papua New Guinea.

The extent and quality of the data allowed us to integrate the trajectories of meteoritic fragments from the airburst site to the surface of the ocean. Or, more scientifically, the initial velocity of 44.8 kilometers per second with an azimuth angle of 285.6 degrees and an elevation angle of 26.8 degrees tells us so. Or, more scientifically yet, it does so because we can also calculate IM1's deceleration timescale based on the ram pressure on an object with a well-inferred radius, speed, and density, as well as knowledge of the air's density at given altitudes. This helps us to arrive at a minimum terminal velocity, based on informed approximations of the mass of the fragments and acceleration due to gravity. Thereafter, and calculating in things like ocean surface impact and ocean current drifts, we know with relatively high confidence where we must go to lay hands on what remains of IM1.

Plate activity along with the more concerning ocean currents and composition of the seafloor were among a slew of considerations on my mind when, on September 16, 2022, I wrote, "Wonderful news! We just received funding for the Galileo Project's expedition to scoop up the fragments of the first interstellar meteor." By my count, it was a moment eight years in the making. IM1 had been detected in 2014, it had been identified as interstellar in 2019, its discovery as an interstellar meteor was confirmed in 2022, and before that year was up we had the funding necessary to undertake the expedition to scoop its fragments up. Having secured the approximately $1.5 million needed to go and retrieve some of IM1's fragments, it was possible that in the near future humans would hold material from a meter-sized object known to have been created outside of our Solar system.

In one context, I understand that putting dollar signs to scientific undertakings can feel crass, and often (sometimes rightfully) raises the question, "Isn't there better uses for such sums of money given the problems humans face right now, right here on host planet Earth?" To this real and significant debate I will offer here just one consideration. The species invoked in the thought experiment of Chapter 1, a species organized and prepared to discover unequivocal evidence of another technological civilization's existence, spends less time arguing over these points. Perhaps, it is to be hoped, it is because they have resolved enough of their host planet's problems that they are able to consider scientifically their interstellar future.

To me, the prospect of spending one and a half million dollars to hold interstellar spherical micrometeorites is amazing. To hold in our collective hands an object not of our Solar system, a product of a distant part of the Universe's formation, and to study its properties would be a first-ever scientific accomplishment. To study it in our laboratories so that we can begin to explain its composition, and compare it to local Solar system meteors, will help answer questions both cosmological and practical. The chance that IM1 just might yield data that allow us to reach scientifically plausible hypotheses about its extraterrestrial manufacture could, should, change everything.

Our plan is to mow the ocean floor with a magnetic sled of our own design and manufacture. The plans for the prototype imagine a 2-meter-long, 1-meter-wide, and 20-centimeter-tall sled weighing in at about 55 kilograms. It will be built of polyethylene with an ultra-high molecular weight and sport 20-centimeter-tall steel runners on each side. Along with the magnets, it will host cameras and lights, the latter to aid the expected seven sled operators on the ship. This device will be towed along the seabed, in back-and-forth trips within the 10-by-10-kilometer grid, seeking IM1's ferromagnetic meteorite fragments, and will drag a mesh net behind it capable of collecting nonmagnetic meteor pieces. We contracted a ship built for scientific research and ocean recovery, which can house both the sled operators and a small team of scientists who can study what the sled captures in real time. All of which will happen slowly: the tow speed will be about 3.3 kilometers per hour, resulting in each 10-kilometer run exploring an area of 10 kilometers by 1 meter.

Much as you might empty a lawn mower's bag of its cut grass, the sled will be recovered after each run and any particles attached to its magnets will be removed. With a lawn mower you set the level at which you want your grass cut. With an interstellar meteorite magnet sled, you estimate how deep you need to run it through the ocean floor. For the IM1 expedition, that will be set at about a few centimeters. Fortunately, we know this area of seabed is mostly composed of calcareous ooze and that the sedimentation rate is on the order of millimeters per thousand years. The math works as follows. In seven survey days, adding in turns and recovery times, we expect to have about 126 effective towing hours in the field or near to forty 10-kilometer passes. If, as is being considered, we add a mesh net that would follow the sled, and catch larger particles, more time would be needed. For now, we envision a trip that includes ten survey days.

We also might get lucky. Given IM1's airburst explosion energy, we can estimate the energy per unit area delivered to the seafloor 20.4 kilometers below the explosion. Were you so unfortunate to have been immediately near the site of impact, you would have been privileged to see steam rising from the site. Beautiful, but also deadly. An umbrella would not have been very effective protection from the resulting hot, radiated rain. For the expedition, however, the power of the explosions proved to be a positive.

They allowed us to narrow down the search area by more than an order of magnitude, improving considerably the expedition's prospects for success. The IM1 coordinates published by the United States government defined the fireball location to within a square region ten kilometers on each side. Searching an area of that extent over a ten-day expedition increases the odds of coming up empty. To my delight, however, Amir and I found that the blast wave from the meteor explosion generated a high-quality signal in a seismometer located at Manus Island in Papua New Guinea, nearly a hundred kilometers away from the meteor explosion site. The sound signal includes two broad peaks separated by about four minutes, each lasting for about a minute. At face value, this signal structure is surprising since the explosion lasted less than a second. How did the sound signal obtain this extended temporal shape?

During an early morning in December 2022, I had a eureka moment. Within less than a kilometer from the explosion, the initial shock wave weakens to a sound signal. That signal continues to propagate through air, water, and ground. The speed of sound in air is the slowest by an order of magnitude, being only a third of a kilometer per second. The different paths through the different media allow independent geometric constraints visible in the two peaks registered by the seismometer and separated by about four minutes.

The first peak begins with a sound path that goes through air until it hits the ocean's surface, and then through water and the ground until registering with the seismometer. The shortest path through the slowest medium, air, goes directly from the explosion to the seismometer and defines the beginning of the second registered peak.

Simple geometric models for the propagation of sound waves from the explosion and their reflection off the ocean surface provided many more constraints and allowed us to determine the explosion elevation and distance to within a kilometer. The expedition will start under the explosion coordinates and scan the direction of motion of the meteor, expecting the fragment size to increase in size along it. This discovery will guide our treasure hunt for interstellar fragments at the bottom of the Pacific Ocean.

As the physicist Richard Feynman noted, there is a pleasure of finding things out. In the context of localizing the first interstellar meteor, this pleasure is accompanied by the practical benefit of accomplishing the expedition goal. There is nothing more thrilling than finding an explanation to data that fits an over-constrained model.

Our confidence in discovering remnants of IM1 is in part because what we're proposing to do isn't novel. It has been done before, albeit only once before and in the recovery of a meteor that originated in our Solar system. On March 7, 2018, seismographs both on shore and on a submarine, along with weather radar imagery and a weather buoy, identified the spot where a meteorite had landed in the Pacific Ocean off the coast of Washington State. A research vessel was nearby and so by July of that year a search for meteorites on the seafloor commenced.

Dr. Marc Fries, a scientist with NASA's Astromaterials Research and Exploration Science Division, used a sled to collect micrometeorite-sized melted spherules that had come to rest 350 feet below the surface, lodged in mud and sand. Two of the fragments were between two or three millimeters in diameter, with a single millimeter being about the height of ten stacked pieces of paper. Marc found most of the micrometeorites by painstakingly running a bird's feather over collected samples, the most efficient method he could find to isolate the meteor fragments.

Fortunately for the Galileo Project, Marc had been lending his expertise to the expedition to mow up IM1. It is hoped, however, that our instrumentation has improved from his feather. What is hoped for more, however, is that we will find something larger than microspherules.

What is it that we hope to find? Something discernably of extraterrestrial manufacture propelled by artificial means. Scientists are animated by hope. Science isn't. What I am prepared to find is a meteor of natural origin, likely originating from a deep potential well such as found in the interior of a planetary system, within the orbit of a Mercury-like planet around a Sun-like star. There is also the possibility that we will discover something ambiguous, material of some alloy that, as far as we know, nature doesn't put together.

The ideal scenario, of course, is a piece of an advanced technological device with a button to press. This is admittedly unlikely. First, because naturally occurring things vastly outnumber artificially created things. There are far more naturally occurring things even here on planet Earth than there are manufactured things. Even if we assume that numerous extraterrestrial civilizations have existed over the Universe's nearly 14-billion-year history, and even if we allow that a few of them were not just interstellar but generous to careless with the extent of their manufacturing, there will still be far more rocks out there than AI probes. Also, just as on the streets of any terrestrial city it is far more likely that you will encounter useless, intentionally discarded trash than a free, mint-condition mobile phone or laptop, so is it far more likely that any extraterrestrial artifacts we encounter will be long-discarded, long-forgotten flotsam and jetsam. As such, we shouldn't expect them to come to rest on the ocean floor intact.

Humanity has never bothered to send into interstellar space a craft deliberately designed to survive contact with an Earth-like atmosphere. If Voyagers 1 and 2 survive their interstellar journeys only to slam into an Earth-like atmosphere and come to rest on the ocean floor of a distant exoplanet, it will require a technologically and scientifically advanced civilization to track down their remnants and reverse engineer from the wreckage anything approaching NASA's original intentions. We can also surmise that it is unlikely most other civilizations capable of sending out interstellar craft would have built them to withstand contact with Earth's atmosphere. Humanity has built only a few dozen things designed to withstand re-entry into our atmosphere, and not all of them have survived the encounter. Most of what we manufacture is manufactured to the lowest possible tolerance of routine use that market demand and planned obsolescence requires. And as anyone knows who has ever bought an item that has later been recalled, sometimes even a basic standard of quality is beyond our manufacturers' reach. As unencouraging as these conclusions are, they also lead us to the following wishful thought: if any workable extraterrestrial artifact *is* found following the expedition of IM1, we can safely infer it was intended for us to discover and potentially use.

Now is a useful moment to recall Enrico Fermi's famed query, "Where is everybody?" Given Fermi's demonstrable talents as a physicist, we have tended to hear his questioning the existence of extraterrestrial life and civilizations as the deeply considered insight of a colossus of twentieth-century science. Yet, seven decades ago, when Fermi asked his question, we had limited capability to notice comets and meteoroids, interstellar or not. Caltech's 200-inch Hale Telescope on Palomar Mountain in Southern California, which between 1948 and 1993 remained the world's largest, was its generation of astronomers' most ambitious undertaking. As an instrument of discovery, it has since 1993 been eclipsed many times over, most recently by the James Webb Space Telescope, which doesn't sit atop a small mountain near San Diego, but floats a million miles from Earth. Indeed, Fermi's question, asked in 1950, was akin to an empty-handed fisherman without net, rod, or boat looking at the ocean and asking, "Where are all the fish?" What Fermi, and all who have followed his quip to a dead end, miss is that the absence of discovery very often turns on the effort expended on discovering.

Humanity has a history of struggling to acknowledge its limitations. A 2018 study by the American Automobile Association showed that while 90% of automobile accidents in the United States involve human error, 73% of American drivers consider themselves better-than-average. This finding is so common it has its own name within psychological literature: the better-than-average effect. A 2020 meta-analysis of years of studying the effect concludes what most of us in our private moments might still struggle to admit: there is "a robust tendency for people to perceive themselves as superior compared with their average peer." This effect may be more pronounced among scientists. Which reminds me of the tale told by the German physicist Hans-Peter Dürr about the fisherman who discovered a new law of nature. "All fish are bigger than two inches," the fisherman confidently proclaims—after all, he's fished his whole life and never seen one smaller. Only then is it pointed out to him that the holes in his fishing net are 2 inches wide.

When it comes to seeking useful data, size matters. This is a function of nature rather than of any human biases.

No matter the weather, my morning routine includes a three-mile jog. Living in Massachusetts, "no matter the weather" means that I might encounter frigid temperatures, snowstorms, rainstorms, and (thankfully) a statistical majority of days boasting clear-to-cloudy skies and no significant precipitation. The weather isn't the only variable, however.

The space between the ground and sky is full of life in the form of birds. Their populations change seasonally, but even though their numbers vary, what never changes is the fact that there are always more small birds than large birds. The number of cardinals I will spy over thirty minutes is always going to be greater than the number of hawks, and come spring, the number of robins, blue jays, and chickadees is going to be many magnitudes greater than the number of hawks. And while bald eagles were reintroduced to Massachusetts in 1982, I've yet to see one.

Under the banner of "so many things to do, so finite an amount of time," I have not taken up birding as a hobby. But there is a parallel between bird populations and my day job studying the early Universe, black holes, dark matter, and the possibility of ETC. A generalized rule of nature is that it produces a great many more small things than large things. Just as nature has produced more small birds compared to larger birds here on Earth, interstellar space is likewise full of far more small objects than large ones. If you are a birder in Massachusetts, you will have to work far harder to check off spying a bald eagle in the wild than you will to see a chickadee. But here the parallel starts to break down. For while very few will be impressed by the statement, "I saw a blue jay today!," the world will rightly stop and take notice when we are able to say that we have found and retrieved a small, fragmentary interstellar object.

While nearly all humans have seen small birds, no human has yet found, retrieved, and held an interstellar object or its fragments. This is a function of the quality of our cosmic fishing nets.

Over the last two decades, humanity has identified four interstellar meteoroids. The largest one is 2I/Borisov. Discovered in 2019, it was about five football fields—or half a kilometer—long, which, NASA states, made it "relatively small." By this they meant relative to other comets. To Galileo Project members, it was also relatively uninteresting, for it was a comet that was not much different from those frequently found in the Solar system. The second largest interstellar meteoroid is 'Oumuamua, discovered in 2017; measured at over a hundred meters on a side, it was about a fifth the size of Borisov. In part because 'Oumuamua had no cometary tail, just what it was remains a matter of active debate.

The two smallest interstellar meteors discovered to date are IM1 as we know, and IM2, which was identified in 2022. The second, at about a meter in radius, was another interstellar messenger that Amir and I discovered on close examination of the data captured by CNEOS and confirmed by NASA. IM2, which was detected on March 9, 2017, scattered its fragments in the waters off Portugal. Though IM2 is the larger of the two, at about a meter in size, IM1 is the easier to plan a recovery expedition around ("easier" being a very relative term). The more important point is that within the orbit of the Earth around the Sun there are a million interstellar objects like IM1 and IM2 for every 'Oumuamua-sized object. They are, however, too small to be detectable by their reflection of sunlight by our current instrumentation. They represent invisible fish that are captured once per decade only because the Earth is a tiny fishing net, a billion times smaller in area than the orbit of the Earth around the Sun.

Before 2017, humanity had identified no known interstellar objects. 'Oumuamua was rapidly followed by 2I/Borisov, which was rapidly followed by IM1 and IM2. The latter are, of course, the only interstellar meteors discovered so far. Notable is the fact that IM2 evidenced similarly outlier composition strength in comparison to all known meteors, just as did IM1. Out of the 273 fireballs caused by meteors in the CNEOS catalog, they rank 1 and 3, respectively, in terms of material strength. This implies that interstellar meteors come from a population of material strength characteristically higher than meteors originating from within the Solar system.

How do we explain that? What would account for the fact that the first two interstellar meteors humans discovered exhibit far greater material strength composition when compared to the known population of meteors originating from within the Solar system?

Our first assumption would be that they're not rare. A large percentage of interstellar objects are likely similar to IM1 and IM2 in size and material strength properties. How would such an abundance of stronger meteors be naturally generated? Iron meteors, among the strongest naturally occurring meteors, are created in the sufficiently hot molten cores of planets and large asteroids. Following random collisions, these cores are exposed to the cold vacuum of space and cool to iron meteorites. How many random collisions exposing how many such cores would produce a large enough background population of interstellar objects so that the first two humans happen to discover are of material stronger than 95% of the Solar system's locally generated meteors? In a paper Amir and I published in

2022, this one accepted promptly for publication in *The Astrophysical Journal Letters*, we wrote that "the detections of IM1 and IM2 combined imply that ~2/3 of the mass budget in stars is necessary" to provide the refractory elements, or naturally occurring elements with high melting points (think iron and tungsten), that would naturally produce a population of interstellar meteors of the material strength of IM1 and IM2. "The extraordinary mass budget required to produce interstellar meteors seemingly defies planetary system origins," we concluded, "and suggest some other highly efficient route for creating meter-scale objects made of refectory elements."

We confront, in other words, yet another scientific anomaly at the dawn of our interstellar age. Perhaps we will learn of new natural processes that produce refractory elements. Perhaps on studying fragments of IM1 in our laboratories we will discover qualities about its material composition that encourage us to considered nonnatural origins. Perhaps we will discover IM1 to be an extraterrestrial manufactured artifact. At present, all we are certain of is that when it comes to our interstellar future, there is so much left for humanity to learn.

A sense of the breadth of what confronts the scientifically curious was caught during one week in April 2022. First came Dr. Mozer's confirmation that the velocity estimate of IM1, which Amir had been the first to identify, established the meteor's interstellar trajectory to an almost 100% confidence. Within a day of Mozer's tweet confirming those data, I enjoyed another eureka moment: my research group published a press release announcing the discovery of the farthest object ever then detected by humanity, a galaxy named HD-1. Discovering it required the combined efforts of the Subaru Telescope in Japan, the VISTA Telescope in Chile, the UK Infrared Telescope in Hawaii, and the Spitzer Space Telescope orbiting Earth. Based on a wealth of data, this galaxy is inferred to have a redshift of 13, implying that it emitted its light only 300 million years after the Big Bang. This made HD-1 at that time not just the most remote but also the oldest massive galaxy ever detected by then.

Over a single week in April, I therefore had the pleasure of being involved in the discovery of both the closest and the farthest objects from outside our immediate cosmic neighborhood. That pleasure arises almost entirely from the number and nature of the cosmic fishing nets now at astronomers' disposal. The discovery HD-1 was good news for the James Webb telescope, which has since commenced its cataloging of numerous early massive galaxies. The discovery of IM1 was good news for the Galileo Project, which has since commenced its pursuit of data on UAP and ISO. Both HD-1 and IM1 underscore that there are many more of both massive galaxies and interstellar objects awaiting our attention and discovery. They also capture the wonderous breadth of this new interstellar era of humanity.

The light HD-1 emitted took 13.5 billion years to reach us. This means, of course, that the source of the light is even farther because of cosmic expansion. Just what that rapidly receding source is, is a mystery. The light from HD-1 could have been emitted by a burst of star formation, in which case the stars should have been mostly massive. As argued through my own theoretical studies decades ago, this is expected for the first generation of stars that we commonly label Population III stars. But the light could also have been emitted by a black hole weighing 100 million Suns.

As the earliest supermassive black hole, HD-1 would represent a giant baby in the delivery room of the early Universe. It would break the highest quasar redshift on record by almost a factor of two, a remarkable feat. This would also allow us to surmise a few things about its early life. In the short time available for its growth right after the Big Bang, it must have started with a massive seed and increased its mass by eating matter from its immediate environment at an unprecedented rate. Such a result would not only be surprising, proving once again that nature is more imaginative than twenty-first-century human science, it could also be demonstrated. There are many data for humanity still to gather to make sense of these moments in the early birth of the Universe. If HD-1 is a black hole, we should find X-ray emissions from it. The Chandra X-ray Observatory could detect such X-rays, or not. In the latter case, the emission must originate from massive stars.

Knowing matters all the more because we are now an interstellar civilization. Ascending the ladder of civilizations requires us to know all we can of the shared Universe we are expanding into. All life on Earth is connected immediately to every interstellar object and each rapidly distancing galaxy. All of it shares this common Universe. All of it is connected, traceable to a single Big Bang origin story. From this cosmic perspective, there are no in-groups and out-groups, there is no partisanship. Humanity, in all its immature interstellar infancy, can already imagine sentient AI and 3D printers capable of producing anything, including life. All of which points to the fact that among the civilizations of the Universe, there are only finite circumstances, best measured in relativist time.

Humans currently race a very terrestrial clock. From 1950 to the present, we have managed to launch five spacecraft destined to enter interstellar space. We have managed to identify four interstellar objects. And we have managed to increase the parts per million carbon dioxide level from approximately 310 to approximately 420. While Earth's climate has changed throughout the planet's long history, which has included eight cycles of ice ages and warmer periods, those changes are clearly attributable to small variations in Earth's orbit. Today's climate change is different because today's climate change is unambiguously due to human activity. All of the following are confirmable by robustly accumulated data over decades: our oceans are getting warmer, our ice sheets are shrinking, our glaciers are retreating, extreme weather events are increasing, and sea level is rising. A simple calculation about humanity's future prospects as an interstellar species is how many interstellar craft, and with what abilities, are we likely to send out before Earth's temperatures rise to levels that initiate catastrophic consequences.

It is my hope that the expedition to Papua New Guinea will reveal that IM1 was made out of an artificial alloy. The moment such a finding is confirmed, humanity would begin to enjoy a realization that will change our technological self-esteem, our aspirations for space, and our long-term agenda. Given our shared circumstances as a *D-class* civilization, we should all share the same hope. Those same circumstances demand we share it scientifically.

My critics have declared that Professor Loeb sees artificially created extraterrestrial debris wherever he looks. Nothing could be further from the truth. What I do is allow for a hypothesis that matches available data, and thereafter I seek better data to test, confirm, or reject my or other competing hypotheses. That I do so with human worries and wishes only means I am human. Another aspect of being human, however, is the wherewithal to allow science to direct, mediate, and temper my worries and wishes.

While seeking partners to help in the recovery of IM1, I had numerous opportunities to describe the vision of the expedition to the waters off Papua New Guinea. One such arose at a conference in New York City. Afterward, over dinner with Kevin Conrad, a Papua New Guinean businessman, I laid out the scientific case for seeking the remnants of IM1. It is possible that what we will find is an interstellar rock of unique properties. It is possible that what we will find is only plausibly of extraterrestrial manufacture. And if it is clearly of extraterrestrial origin, it is highly likely that we will not be able to make sense of or translate its purpose or message. What it does and what it might mean to tell us would be written in a language inaccessible to us. Smiling, Conrad said that, if this were true, then IM1 had landed in the right part of the planet.

He told me the fascinating fact that there are nearly 850 languages spoken in Papua New Guinea alone, a country with only 7.6 million people. This makes it the most linguistically diverse country on Earth. Considering the possibility that our expedition will discover that IM1 carries a message in a technological bottle lying on the ocean floor, I granted that he was right. Just maybe this message would bring yet another new language to Papua New Guinea, this time from an interstellar origin. And maybe, just maybe, the recovery of the first interstellar meteor ever identified as such would further encourage a human civilization commencing a common interstellar future to act as one.

5

LEAVING EARTH

IF A SPACESHIP LANDED in your backyard, would you board it?

My answer has long been a resounding yes. And for many years, on hearing my answer, my wife, Dr. Ofrit Liviatan, replied, "If you decide to go ahead, I want you to do two things before you board the spaceship: please leave the car keys with me and ask them not to ruin the lawn at liftoff." Recently, however, she updated her answer: "If you go ahead, please turn off the lights. I will join you." With a sigh of relief, I understood with that answer she was agreeing to step into humanity's interstellar future.

There are few certainties in life, but one is that someday humanity, if it wants to persist, will need to leave this planet. The cosmic clock is ticking. In a few billion years, the Sun will begin to die and life as we know it on Earth will perish. For humanity, the clock to a partial or complete extinction is likely ticking far faster. No matter how much I wish for our civilization to give up its *D-class* habits and assume those of a *C-class* civilization, I have my doubts. I also doubt that any extraterrestrial civilization is going to land on Earth and offer to give us a ride off. The likelihood that all UAP are explainable without recourse to extraterrestrial technological civilizations is *the reason* we must be certain. When it comes to getting off a host planet, it isn't the hope that kills you so much as it is overreliance on a falsifiable hope.

The search for UAP and ETC goes immediately to human considerations of how some of us transition to a *B-class* civilization. Recall, a *D-class* civilization is degrading its planet's ability to sustain life, and a *C-class* civilization is optimizing its planet's ability to sustain life. A *B-class* civilization is managing to sustain some life off-planet and independent of its host star. As is always the case in considering the possibility of extraterrestrial technologically advanced civilizations, we most often work with transitive logic: if they did it, we can do it; if we can do it, they could have done it. Our impediments were once theirs. Discovering how they overcame them could be one means for how humanity overcomes its stuck points. Imagining how they overcame them is another.

As of 2022, we are a very young interstellar civilization able to boast of only five human-manufactured probes heading into interstellar: Voyagers 1 and 2, Pioneers 10 and 11, and New Horizons. We remain ignorant about our Solar system, especially as to whether any other planet or moon sustained or sustains life, whether any interstellar artifact rests below the surfaces of the system's planets and moons, whether any interstellar artifact is accidentally or intentionally passing through our planetary system or orbiting Earth. Even at humanity's current level of technological prowess, we know how to send out more and better interstellar craft, how to search for life on nearby planets and moons, how to send a very small, very fast craft to briefly survey an exoplanet over four light-years away.

None of these are matters of imagination, but matters of determination, and a few—seeking evidence of water or life on Mars, Venus, or Titan, for example—are underway. If human civilization can, another civilization could, and the odds are great that they did it for longer and more skillfully. The discovery of a single, even fragmentary and defunct extraterrestrial artifact would vastly accelerate humanity's interstellar future. Until we do, approaching the moment of that discovery as a matter of when, not if, is our best and most promising means to ascend the ladder of civilizations.

THE TYRANNY OF ROCKETS

Currently, all of humanity lives under the thumb of a planetary-wide technological tyrant: our rocketry. That our civilization is now reliant on propellant-powered spacecraft is a collective choke point to our progress. Understanding the tyranny, and a few of the promising paths by which we overcome it, braids several threads of our interstellar future. If any of the UAP admitted to exist by the Pentagon are conclusively identified to be of extraterrestrial technological manufacture, it is a near certainty that they are not traveling by way of propellant technology, and it is only the scientist in me that inserts the word *near* in this sentence. Of all the hypotheses that could explain reported UAP behavior, that they are craft manufactured by civilizations that have mastered propellant technologies in ways to permit the UAP's purported velocity and maneuverability is the least likely. If any ETC artifact, whole or fragmentary, is discoverable by us, it is a near certainty that it was manufactured by a civilization untethered to propellant rocketry. To understand why is to understand one of humanity's most glaring challenges in these early years of our interstellar future: the tyranny of the rocket equation.

Credit for grasping rocketry's despotism goes to the Russian (and eventually Soviet) scientist Konstantin Tsiolkovsky, who in 1897 came up with the mathematical equation that explains the acceleration of a vehicle that uses the thrust generated by ejecting burnt gases through its exhaust.

If you have ever traveled to Cape Canaveral, Florida, to watch a NASA rocket launch, or have just seen pictures of crowds safely observing a liftoff, you'll notice that just about everyone in attendance is watching the rocket. Understandable, but doing so misses a necessary big thing, namely Earth. The only reason the rocket rises is because it has something to push off of, in this case our planet.

Propellant rockets need something to push against, spending energy when they do. The closer you are to an object's gravity well, in this case the center of Earth, or where the greatest amount of energy is required to move a mass, the more energy you need. You expend far more energy achieving liftoff from Earth than you do achieving liftoff from the Moon because the former's gravity, in meters per second squared, is 9.8 compared to the Moon's 1.6.

Once a rocket has obtained low Earth orbit (LEO), far less energy is needed to get it to move. But, to get any rocket in space to move, and move ever faster, it must still increase its exhaust velocity, or the speed at which exhaust leaves a rocket engine relative to the rocket. The vast, vast majority of human-built rockets depend on chemical propellants, which have different exhaust velocities. Fluorine-hydrogen, for example, provides a factor of 300 to 385 in the mass ratio that the fuel is capable of lifting from the ground per second, while hydrazine provides 160 to 190. All of them are, by design, combustible. The space shuttle Challenger, which tragically exploded in 1986, was propelled by liquid oxygen, a volatile but light fuel.

A principle of global momentum conservation applies to Tsiolkovsky's insight that the fuel mass required to propel a rocket grows exponentially with the increase in its desired terminal speed. This tyranny of the rocket equation stems from the fact that the rocket carries its fuel for the ride and so the momentum thrust delivered to its payload must be shared with the unburnt fuel. To increase the speed of your rocket, you need to exponentially increase the amount of fuel your rocket must carry. Fuel mass a million times the payload mass implies that the terminal speed of the payload cannot exceed fourteen times the ejection speed of gas through the exhaust. Fuel mass a billion times the payload mass implies a factor of twenty. The characteristic exhaust speed of chemical propellants, a few kilometers per second, has limited all rocket launches, from Sputnik 1 to New Horizons, up to terminal speeds of several tens of kilometers per second.

Travel time from Earth to Mars on a rocket going approximately 30 kilometers per second is approximately 785 hours at closest separation. From Earth to Jupiter, 6,500 hours at closest separation. And from Earth to Pluto, nearly 50,000 hours at closest separation. And from Earth to the somewhat nebulous boundary of the Oort cloud of Solar system building blocks that marks the beginning of interstellar space, around 20 trillion kilometers, which would take 180 million hours or 20,000 years on that steady-speed rocket. At 4.4 light-years away from Earth, Proxima Centauri is 44,000 human years away by our conventional rocketry.

At the risk of sounding glib, interstellar humanity has a long way to go. Less appreciated by humans is that with the passage of time, it is only getting longer to venture beyond our galaxy.

During the first half of the history of the Universe, the cosmic expansion rate slowed down owing to the attractive gravity imparted by radiation and matter. During the second half, however, radiation and matter were diluted so much that the repulsive force of dark energy in the vacuum of space dominated and cosmic expansion accelerated. The actual mass budget of the Universe is currently dominated by the vacuum, and its repulsive force is causing galaxies to recede away from each other at an ever-increasing speed. Within ten billion years, the cosmic acceleration promises to separate galaxies up to the speed of light. Which means, we will then not see light emanating from these distant, fast-moving galaxies. At that time, we would exit from our event horizon just like a light bulb falling into a black hole. If our interstellar future is to include visiting neighbors, we need to get busy besting the tyranny of propellant rocketry.

One way we might do it, via lightsail technology, has already been deployed. In June 2015, the Planetary Society launched the crowdsource-funded LightSail 1, humanity's first such propelled spacecraft. Just as anticipated by the Breakthrough StarShot Initiative, the craft was small—4.5 kilograms in weight with dimensions of 34 x 10 x 10 centimeters. That's about the size of a large rectangular box of spaghetti. Its solar sail, made of Mylar, a reflective polyester film, is considerably larger, with a cross section measuring 32 meters squared. A proof-of-concept mission, LightSail 1 successfully deployed and is currently orbiting Earth. LightSail 2, about the size of a loaf of bread with a sail about the size of a boxing ring, successfully deployed in July 2019. Unlike its predecessor, LightSail 2 is the first spacecraft to demonstrate controlled solar sailing.

While the payloads of lightsail craft will remain modest for the near future, they are functionally improving, and we can nearly guarantee human civilization's ability to seed space with them, if we choose to.

Another design for spacecraft, still in the conceptual stage, is gravitational propulsion. This method depends on our ability to propel a craft by manipulating the cosmic dark energy that makes up most of the vacuum of space. According to Einstein's equations of general relativity, gravity is sourced not only by mass or its equivalent energy, but also by pressure. In a uniform medium—like the surface of the Earth—the strength of gravity is dictated by the energy density plus three times the pressure. For radiation, or hot matter, the pressure is a third of the energy density, and so gravity is doubled compared to cold matter of the same energy density. But the vacuum is qualitatively different from matter and radiation. It has a negative pressure with a magnitude that is equal to its energy density. The result, repulsive gravity, is driving galaxies ever faster and farther away from each other. In theory it could also propel a spacecraft.

So far, cosmologists have no clue about the nature of cosmic dark energy. And for us to propel ourselves by manipulating dark energy, we would first need to excavate or engineer a substance, dark matter, we do not understand. We can hold to a hope, however, that exo-cosmologists have figured it out. If such craft exist, they would be harder to spot, because they would move without anything coming out of their exhaust. But it should be possible to identify them because they would still interact with light, air, and water. Their passage through our atmosphere at extreme speeds, for example, would still produce a fireball.

But even if the negative gravity of the vacuum cannot be used for propulsion, normal attractive gravity could be harnessed through gravitational assists. We already used our Solar system's optimal gravitational slingshot to speed Voyagers 1 and 2 out of the Solar system. Their launches were planned to take advantage of an ideal alignment of the four outer planets, which occurs every 175 years, picking up speed by way of gravitational assistance as they passed Jupiter and Saturn. A tenth of the speed of light is accessible with certain hypothesized gravitational slingshots, such as one composed of two neutron stars in a tight orbit. Black hole pairs could do even better. Maneuvers such as these, however, would be engineering feats of extraterrestrial civilizations or of far distant generations of humans.

Yet another possibility that also doesn't rely on the negative gravity of the vacuum, or, for that matter, on dark energy or matter at all, is rocketry engineered to take advantage of Modified Newtonian Dynamics. In 1983, the physicist Moti Milgrom suggested that the missing matter Fritz Zwicky called "dark" might not be missing at all. Zwicky, recall, started with his observing stars rotating in a galaxy at speeds inconsistent with the mass visible to his instruments. His solution was to theorize more, but invisible, mass. Milgrom asked, Can the observed speed at which the stars revolve be explained without theorizing dark matter? Yes, if Sir Isaac Newton's law of gravity, which applies consistently on Earth, was theorized to apply differently at the low accelerations experienced in the outer edges of a galaxy.

Leaping over the math and the particulars of Milgrom's theory, know that four decades after Milgrom presented it, his low acceleration law of Modified Newtonian Dynamics (MOND) still agrees with most available data on galaxies of all sizes and shapes. While most astrophysicists still find Zwicky's dark matter more helpful in explaining the observable Universe, MOND holds out one possible escape from the propellant rockets' tyranny.

If the modified inertia interpretation of MOND holds, it would allow rockets to traverse intergalactic distances by consuming modest fuel mass. For rockets launched at low accelerations, Tsiolkovsky's tyrannical rocket equation is suppressed. This would offer the opportunity for a rocket to reach a terminal speed that is a hundred times larger than the exhaust speed if the acceleration is a hundred times smaller than MOND's threshold acceleration. A chemical rocket with a fuel mass comparable to the payload mass could reach speeds of hundreds of kilometers per second, ten times faster than all space rockets launched from Earth so far.

You don't have to take Milgrom's or my word for it. The exciting reality of the modified rocket equation can be tested by launching our own low-acceleration rocket or by finding low-acceleration rockets arriving in our cosmic neighborhood from great distances. Either option, and humanity is encouraged to pursue both, requires a civilization interested enough to try.

We now approach a dividing line, one of many. There was humanity before 'Oumuamua and humanity after. Throughout the year 2017, the two were nearly indistinguishable. Before 'Oumuamua's discovery in October, we were a civilization considering the possibility of interstellar objects, considering the possibility of UAP and ETC. By November, we were a civilization with conclusive proof of interstellar objects, one of which demonstrated wildly outlier properties in comparison to all known local space rocks, properties most easily explained by manufacturing, with the conceptual difficulty being humans couldn't have manufactured 'Oumuamua. By 2022, the existence, but not the nature, of UAP was robustly confirmed by the United States Department of Defense. In that same year, we identified more interstellar objects with outlier properties, locating one of them off Papua New Guinea. And as we know, over this span of years, from 2017 to 2022, a rising number of organizations, governmental and private, began to actively organize and deploy instrumentation to improve the data collected on UAP and interstellar objects with sufficiently outlier properties to raise questions of their origins, natural or artificial.

In spite of all this, the world and human civilization as I write this paragraph doesn't look materially different to the world and human civilization as it existed in early 2017. The reason is simple. The Universe, and our planet within it, hasn't changed at all. This is true even as dramatic, civilization-altering events occur. Most such events, however, are noted by only a few. That something might happen that galvanizes everyone all at once is possible, but remote. Just as it is exceedingly unlikely any extraterrestrial craft will land in my, or your, backyard. If our civilization is to persist, if it is to continue to mature and accomplish ever more things, some percent of our sentient intelligence must get off-planet, must learn to thrive without reliance on the Sun. At this inflection point in our interstellar future, it is a game of numbers. Some portion of humanity is actively seeking means to ascend the ladder of cosmic civilizations, whether we do that through the efficiency of an extraterrestrial assist or by our own plodding, homegrown efforts. My wife and I number as among that portion of humanity. The question is, Has a sufficient portion of humanity come around to understanding it is navigating itself toward an interstellar future? What would be optimistic, hopeful signals that it has? What wouldn't?

ESCAPE TO REALITY

Our current technology allows human-manufactured artifacts to travel interstellar space. It allows us to probe near planets and asteroids. To plan returning astronauts to the Moon and more robotic expeditions to Mars. We already have a propellant technology—the solar sail—that allows very small craft to escape the tyranny of chemical-propelled rockets. Other methods of slipping the bonds of that tyranny are theoretically plausible, even probable. And we already have, though only just barely, an interstellar technological footprint. Our greatest obstacle is not ability—it is determination.

I propose that the canary in the coal mine of human civilization is our love of evidence. It requires our collective willingness to lay the head of wishful thinking under the guillotine of data. Imagined realities are addictive. They satisfy deeply held emotional needs. Our submission to experimental data gathered by instruments feels impersonal, and in science that is by design. What is certain, however, is that no civilization lived a long, productive interstellar existence by following a path of wishful thinking. Real-world evidence virtuously dispels misleading wishes without prejudice. The astrophysicist Henry Norris Russell dissuaded the young student Cecilia Payne-Gaposchkin from publicly declaring her theory that the Sun is made mostly of hydrogen, until his own years of research proved her right. Albert Einstein famously resisted quantum mechanical entanglement, yet in 2022 the Nobel Prize in Physics was shared by three pioneers who demonstrated experimentally the reality of quantum entanglement.

Reality has repeatedly proven to be far weirder than human intelligence can imagine. What it has always stubbornly remained, however, is implacably real.

I watch with curiosity and mounting concern contemporary talk of the pleasures and benefits that human civilization will enjoy from virtual reality. Its most evangelical admirers promise the dawn of a new, wondrous era for humanity, one defined by the attainments and opportunities enjoyed via digital avatars. Based on my experience, as a scientist, a parent, and a human, virtual reality is not new and its current incarnation—goggles worn to transport you to computer-generated 3D experiences—is no less dangerous to civilization than delusion-inducing drugs.

In virtual reality you can hop on an alien craft, visit distant exoplanets, enjoy first and repeated encounters with extraterrestrials, and hyperjump among multiverses. What you cannot do is interact with reality, the avoidance of which is among the longest-lived siren songs of our civilization.

To cope with reality, humans have long turned to illusions. We wear wigs and platform shoes, and sit in darkened theaters to watch fairy tales performed by actors wearing wigs and platform shoes. We lap up elaborate wartime deceptions, conspiracy theories, and casual lies, and flock to political, sporting, and religious spectacles. But for me, the most insidious example of our ability to self-deceive is when scientists fudge ugly data to protect a beautiful hypothesis. When you survey the written record of our collective past, it is arguable that humans have spent more of their history in virtual reality than reality, despite the fact that reality is shared by everyone.

The problem is that reality tends not to flatter human egos. And therein lies the danger of the metaverse and 3D computer-generated worlds. Tailored to match our wishful thinking, virtual reality lures us away from the humility required to gain new scientific knowledge, to spur us to the hard work of gathering evidence and using it to understand the reality we all share. Sometimes this reality is more imaginative than we are, so studying it is a learning experience. Most of our scientific learning stems from facts that contradict our preconceptions. Even though such conflicts between data and prejudice do not flatter our reputation as experts or the adults in the room who can anticipate whatever comes along, they are essential for shaping improvements in our understanding of reality.

The centuries it took humanity to accept the reality of heliocentricity is a well-told story. It's an arc that dates from Nicolaus Copernicus's 1543 deathbed publication of the idea through Galileo's house arrest for empirically proving it to eventually Edwin Hubble in the early twentieth century removing all doubt of the Earth's, the Sun's, and the Solar system's insignificance within the wider frame of the cosmos. Similarly, it took decades for Albert Einstein's theory of general relativity to be accepted. First published in 1915, and empirically confirmed by Arthur Eddington in 1919, it remained the stuff of fears and rumormongering for years after. In 1931, a booklet entitled *101 Authors against Einstein* collected the criticisms of leading academics of the era. And a decade earlier, Einstein wrote in a letter to a friend, "The world is a strange madhouse. Currently every coachman and every waiter is debating whether relativity theory is correct. Belief in this matter depends on political party affiliation."

There is comfort in the fact that while acceptance of heliocentrism took centuries, acceptance of general relativity took decades. And its application to improving humanity was similarly swift. The trip length estimated by the GPS app that is directing your car to your desired address is dependent on calculations that take the relativity of spacetime for granted. We should not, however, become too comfortable, too complacent. Einstein's quote works as well, perhaps even better, when you re-read it substituting climate change for relativity theory.

Our civilization's strange madhouse of political debates had over the reality of human-made climate change, despite the overwhelming and increasing evidence for human-made climate change, allows us to return to the tyranny of the rocket equation. The problem of rocketry is surmountable; the greater tyranny isn't technological but human-made. Both immediately touch on our interstellar future. A civilization capable of behaviors that render its home planet uninhabitable is badly in need of other planetary options. A civilization dependent on chemical propellant rocketry has far, far fewer options. The longer we spend confronting these challenges principally from within the madhouse of our terrestrial concerns and cultures, using a language of human ego and historical limitations, the more likely our only-just-begun interstellar future ends prematurely.

The vocabulary that interstellar humanity needs is scientific. In a very real sense, it isn't new at all. We live daily within the now innumerable benefits accrued by a civilization conversant in science, research laboratories, industrial manufacturing, and sufficient social and political stability. We also live daily within the well understood and poorly understood and still not understood externalities of those benefits. Awkwardly present among the ancient Greeks, the scientific method of observation and experimentation became the risky language of Sir Isaac Newton and Galileo Galilei and then became the presumptive language of humanity's rapid economic rise of the last two centuries, even if only spoken fluently by a select few. Science is the lingua franca of the Universe, and the sooner the same is true of humanity the brighter our prospects.

I do not stand with Karl Popper, the philosopher of science, in the belief that all of science devolves to the relentless falsifying of reigning hypotheses. Nor do I believe that Thomas Kuhn, the historian of science, was correct in suggesting science is more like governing fashions, what he called paradigms. Rather there is, in the words of the philosopher of science Michael Strevens, the iron rule of explanation: empirical testing either sustains a hypothesis, or it falsifies it. This accepts that what is empirical, experiences with reality, will be both stable and broadly shared.

Like Galileo's critics, you can ignore the evidence. Like Ptolemy's defenders, you can add layers and layers of epicycles in an effort to sustain a hypothesis. But eventually, empirical testing will falsify Ptolemy and corroborate Galileo. Unless, of course, we use culture or technology to inoculate ourselves from the disconfirmation of a beloved hypothesis.

As soon as I heard about the touted promise of computer-generated, three-dimensional virtual worlds, and the most avid promoters who envision virtual lives and virtual commerce eclipsing the real, it occurred to me that civilizations that swoon to such an attraction solve Fermi's paradox. The answer to the question, "Where is everybody?" might be a disappointing one: they are out there all right, it's just that they are hooked up to virtual reality, where you cannot only be a bored monkey, you can also be an ageless beauty capable of piloting a rocket to Mars circling a geocentric Earth. What you would not do is engage with the actual Universe that all sentient intelligence share.

The dedication to a virtual storyline is particularly addicting for any community of people who are not seeking evidence that challenges and disputes what they hold to be true. This not only applies to political, philosophical, and religious beliefs, but also to scientists who divorced their deliberations from the need for experimental verification, such as string theorists within the mainstream of theoretical physics over the past half century. And wishful thinking about colonizing Mars will not solve the fact that energetic particles bombard that planet and would be a death sentence for terrestrial life within a few years of it taking up residency on the Martian surface. Against such errors, we have the best check to human ego that humanity has ever discovered, the scientific method.

The scientific process can be thought of as Darwinian natural selection of the fittest idea when put to the test of the empirical constraints, gathered through experiments or observation.

It is an approach that applies not just to work done in laboratories but to our daily life. When we anticipate a birthday present, we imagine many possibilities. When we spy the box in which the gift has been wrapped, we reduce those possibilities to those that fit the size of the box. By shaking the box, we can limit the range of possibilities further by attending to the sound that the gift makes as it rattles, or doesn't. Ultimately, by tearing the wrapping paper off and opening the box we remove all doubts about its contents. The often-quoted adage of Sherlock Holmes runs, "When you have eliminated all which is impossible then whatever remains, however improbable, must be the truth." This can boil down to, you wanted socks but got a book. Or, something more magnificent. You wanted to be affirmed as the being at the center of the Universe, but instead learned that you are a run-of-the-mill creature on an unremarkable planet in an unremarkable Solar system, one among billions.

Much like shaking a wrapped birthday present, the scientific understanding of the world goes through stages of gathering more empirical constraints and removing uncertainty about the validity of imagined theoretical possibilities. Most of the time science appears to be what it is, a work in progress. An island of knowledge in an ocean of ignorance. This is true not just because much of the time the evidence scientists are considering is not good enough to isolate a single, ever more confirmed hypothesis. It is also true because the more a hypothesis is confirmed by ever more redundant empirical evidence, the less interesting it becomes to science and to humanity. The book remains a book; it will not become a pair of socks; move on.

Sometimes the empirical world surprises us by showing evidence for unimagined ideas. An example is quantum mechanics, which was not anticipated until it was discovered unexpectedly by experiments. Even after its discovery, prominent scientists like Albert Einstein had difficulties accepting the revolutionary quantum nature of reality, which he called "spooky action at a distance" and today is termed quantum entanglement. Put too simply, the measurement of, for example, the spin of one entangled particle predicts the measurement of the other particle's spin, no matter how far apart the two entangled particles are. This apparent violation of local realism is in fact human ignorance encountering reality. A similar experience might await our discovering life as we do not know it. Perhaps the first hints of that would stem from encounters with interstellar objects that do not look like anything we have seen before.

That, of course, has already happened. Among others, former Navy pilot Lieutenant Ryan Graves reports seeing on multiple occasions radar and infrared data of objects exhibiting inexplicable characteristics. He recalled other pilots in his and other squadrons of F18s on training runs off the coast of Virginia reporting the same. And he has described pilots reporting visual confirmation of a dark or black cube within a translucent sphere sitting stationary, despite offshore winds, as their Navy planes flew past it. After eliminating equipment malfunction and after numerous submitted reports to the appropriate Naval Aviation safety departments, the conclusion drawn is that multiple pilots on multiple occasions reported encountering objects capable of remaining stationary or sustained speeds, and more rarely supersonic speeds, while moving in patterns.

What the Pentagon Report announces is that Lieutenant Graves and others have provided America, and humanity generally, with data. We simply remain ignorant of scientific explanations for what is being reported. The UAP observatories of the Galileo Project are intended to help reduce our ignorance. The efforts of the Interstellar Object (ISO) branch of the Project are intended to anticipate the means of reducing ignorance on encountering something we've never photographed before. Regardless, just the deliberate, scientific preparation for the opportunity to encounter something anomalous, something we've never seen before, can prove the aspiration necessary to take humanity to a new, better, more complete understanding of reality. What if, for example, the desire to understand how a hypothesized extraterrestrial civilization might traverse the Universe leads to humanity ending the tyranny of rockets?

That would follow the very well-established pattern of a civilization that observes, imagines, and experiments. Consider the kite. Like the seismometer, it appeared first in Asia. Records indicate that the first flew about two thousand years ago—my guess would be very shortly after the wind stole its inventor's hat. Refinements, by way of experimentation, soon followed. An ideal kite is an object with the largest possible surface area for its mass. As such, it is exactly the opposite from the edge of a knife, which is sharpened to possess the smallest surface area for its mass. Pursuit of this ideal has launched countless craft, tethered to string and not. Up until the recent past, kites have ordinarily surfed on wind—but now they also surf on light.

That humanity can translate its knowledge of reality into lightsail craft tells us any similarly advanced civilization could do the same. Also, humanity's technological immaturity alerts us to the likelihood that the more advanced a civilization is, the less likely its most easily discoverable artifact will be spacecraft in the taxonomic family and class of every spacecraft ever built by humanity. These, after all, are the least efficient means to ensure a civilization that is reliant on its host planet will persist.

In 2022, I wrote a scientific paper showing that the light emitted by the disk of stars in the Milky Way galaxy can lift thin films of material above the disk midplane. The radiative force pushes the films away from the disk, just like the wind lifts a kite up in the air by overcoming the Earth's gravity. For solid films thinner than a micrometer, the outward radiative push away from the Milky Way disk exceeds the inward gravitational pull toward it. Sunlight would push such films out of the Solar system. The ratio between the Sun's radiative push and gravitational pull is the same everywhere because both forces decline inversely with the square of the distance from the Sun.

Films thinner than a micrometer possess a cross-sectional area per unit mass that is ten thousand times larger than a cloud of free electrons and protons. Such films could be produced naturally by the coagulation of dust particles in the midplane of protoplanetary disks and later blown away from these planetary nurseries like dandelion seeds. But they could also be produced artificially by technological civilizations like, but more advanced, than ours.

If we're to find them, however, we need to go out and look. What we would be seeking is much more likely to reflect facts visible in our real world than wishes conjured in virtual worlds. Indeed, among the concerns I have about the rising enthusiasm for virtual reality is its ability to allow an individual's interests to obscure our civilization's common concerns.

INTERNATIONAL TEAMSHIP

Among President Ronald Reagan's most memorable speeches is one he gave in June 1987 during a visit to Berlin's Brandenburg Gate. It ends on one of his best-remembered quotes:

> General Secretary Gorbachev, if you seek peace, if you seek prosperity for the Soviet Union and Eastern Europe, if you seek liberalization: Come here to this gate! Mr. Gorbachev, open this gate! Mr. Gorbachev, tear down this wall!

Many mark that moment as the end of the Cold War, the end of an era. It was a call to hope, to prosperity, to something better.

Less well remembered and remarked upon is Reagan's so-called alien threat speech that he gave about a month earlier at the United Nations. The very name by which it is now remembered speaks mostly to the stigma that was then, and is still, attached to "Little Green Men." Then, as was true in the 2022 Pentagon Report, a means to mitigate that stigma was to speak in terms of menace:

> Perhaps we need some outside, universal threat to make us recognize [our] common bond. I occasionally think how quickly our differences worldwide would vanish if we were facing an alien threat from outside this world.

The president's purpose at the UN, however, was aligned with his efforts in Berlin. It was to lodge a plea for common terrestrial peace, for swords being turned into plowshares. The Cold War, proxy wars among ideologies, and world wars had defined the twentieth century, and in reaching for the metaphor of an alien threat confronting the entire planet, Reagan was reaching for a frame that might jar UN members toward less belligerence among each other, more cooperation. That a little over thirty years from that speech the planet does indeed confront an existential threat in the form of climate change, and that this scientific fact has occasioned foot-dragging, obfuscation, and conspiracy mongering is sobering.

Decades later, we must now realize our common purpose in turning plowshares into advanced rocketry. If humanity seeks peace, prosperity, and continued existence we do not need a common threat but a common ambition. Underlying the hope that our future will be better than our past is the belief that rational reasoning will prevail over irrational instincts and that the advance of scientific knowledge can become *the dominant feature* of our civilization. Arriving ever closer to this being the case has been, and increasingly can be, our history; it is likely to have been, and remain, ever more the path followed by even more advanced civilizations.

Accepting the arbitration of experimental data gathered in the real world is not just a tenet of science, it is a survival mechanism. Natural selection favors those who adapt to the real world. We tend to forgive the dinosaurs their extinction due to the changed climate caused by the strike of a mountain-sized meteor. With small brains and articulated limbs selected by nature to run and rend rather than grasp and build, they enjoyed dominance over the rest of planetary life without recourse to science or technology. With larger, far more complex brains and well-articulated fingers skilled at grasping and manufacturing, humans could apply technology and science to reality and avoid extinction. If we don't, the fault is ours. There is no need to imagine another sentient intelligence a million years hence judging our civilization's failure to persist as they sift through terrestrial ruins. Another gift of our reasoning, scientific minds is that we can judge our own failure to think and act. We know the steps to take, difficult without question, to slow and even reverse climate change. It will be no one's fault but our own if we fail to take them. Similarly, we already know the steps we would need to take to accelerate our interstellar footprint.

Within centuries, humans could undertake the work necessary so some of us might reside on the Moon. Add a few centuries and we are likely to discover the means of residing below the surface of Mars, or, add or subtract a century or two, on free-floating space platforms. Given these time horizons, it is tempting to conclude that it is premature to contemplate a global policy long before it is required. But with interstellar meteor debris on our ocean floor awaiting discovery, with governmental acknowledgment of numerous sightings of UAP by military pilots, and with the years-old, certain knowledge of the uncertain origins of 'Oumuamua, now is the time for humanity to articulate policies around discovery of artifacts.

Humanity has no precedent. Heretofore, our discoveries have been of a shared reality that routinely indicates that it is unconscious and without teleological purpose. Everywhere human civilization looks, we see physical laws governing visible matter, even down to natural selection's influence over life's biological progression. Purpose, meaning, intent are prospects for sentient intelligence, of which, so far, there is only us. When, after an investment of more than a billion dollars, scientists with the Laser Interferometer Gravitational-Wave Observatory (LIGO) announced in 2016 that by observing a 1.3-billion-year-old collision of two black holes they had confirmed the discovery of gravitational waves, the attentive of the world rejoiced. The ripples in spacetime that Einstein's theories predicted, and that Einstein himself sometimes doubted, had been discovered. After noting that some involved in this discovery had not lived to see its confirmation, Szabolcs Márka, physicist and member of the LIGO team, rightly declared, "It's really a wonderful feeling that you have validated the investment of the tremendous amount of work. And it's not just that you found something, but you gave something to everybody, to the rest of the human race."

What was gifted humanity by the LIGO team was another confirmed piece of the puzzle of the real world.

Why should the search for extraterrestrial artifacts and technosignatures be of popular interest? I am frequently asked this question, and the example used by one interviewer was that of a taxi driver worried about paying rent—Why should the driver care? My answer referred to the law President Joe Biden signed, with bipartisan congressional support, six months after the Office of the Director of National Intelligence delivered the Pentagon Report confirming UAP. It established a federal UAP office to coordinate study of the phenomenon. Interestingly, the congressional mandate for the new UAP office involves a science plan that aims to:

> (1) account for characteristics and performance of unidentified aerial phenomena that exceed the known state of the art in science or technology, including in the areas of propulsion, aerodynamic control, signatures, structures, materials, sensors, countermeasures, weapons, electronics, and power generation; and (2) provide the foundation for potential future investments to replicate any such advanced characteristics and performance.

If taxi drivers held to only the most narrowly defined professional concerns, they would benefit from faster transportation. But human civilization has not, and cannot, hold to such narrowly defined interests. All humans have an immediate stake in the benefits of great leaps forward in science and technology, just as all terrestrial life has a stake in the mitigation of any consequent risks. All life, here and elsewhere, shares two certainties: in the very, very far future, the Solar system will end; in the not-so-distant future, so will we. Between those two facts rest our choices, and arguably the purpose and meaning of our existence.

The only barrier to humanity discovering whether extraterrestrial artifacts exist is our willingness to undertake the search. Similarly to LIGO, we should expect to find extraordinary evidence for ETC only after we invest major funds in the search for it. It would be most appropriate to allocate taxpayer funds to the search for our cosmic neighbors. It would be even more appropriate to pool the resources of all humanity, given the major impact that such a discovery would have on human society.

We already do this within the efficiencies of twenty-first-century civilization. Currently, nationalism keeps company with profit-maximizing capitalism, humanitarianism keeps company with self-interest, and scientific advances across multiple domains, from medicine to rocketry, keep company with innumerable branches of wishful, virtual realities. Consider the behavior of any of humanity's large pharmaceutical corporations. It is noted that only 5% of all their research projects are successful, and only a small fraction of those will reach the consumer market. As a result, most of their scientists spend years on projects that do not improve public health or the companies' bottom lines. This is true also in academia, where most scientific papers do not lead to a new understanding of nature because of invalid assumptions or misleading clues.

Inefficiency is not limited to human activities. It is also common in nature. For example, only a small fraction of the DNA sequence in the human genome is used to make proteins while the rest are without a completely explained function and are colloquially known as junk DNA. Throughout evolution, natural selection has allowed for a strategy of the inefficient redundancy of trial and error. But humanity can do, and has done, better. Improving the efficiency of our actions is as important as lengthening our lifespans. The means are simple. First, we must learn from experience and avoid repeating mistakes. Second, we need to embark on a range of projects to minimize the risks to humanity's longevity that any one of them brings. Third, we should practice agility in adapting to changing circumstances.

In matters of survival of the fittest, whether among species or civilizations, time is the ultimate arbitrator. Humans tend to think short-term as a survival instinct against immediate existential threats. This was brought home to me during the days when the newspapers ran stories reporting on the early images from the James Webb Space Telescope. Those awesome, inspiring images appeared alongside stories of souring global economics, political corruption, and strained supply chains of food, fuel, and goods. Most of us are aware of the real world's immediate risks and opportunities: what is in our bank accounts and refrigerators, the seasonal weather, the near-term political prospects of our nation-state, the ambitions we hold for ourselves, our children, our grandchildren. I worry that this may not be enough. I worry further that some of humanity is adding to our civilization's very real risks through their preference for pursuing virtual opportunities as presented to their avatars. Webb's deep images should inspire us to think not just bigger, but longer, far, far longer.

PART II

6

KNOWLEDGE AND WISDOM

IN SPACE, EVERY COSMIC civilization will put their best technological foot forward. This hopeful thought occurred to me when, shortly after my sixtieth trip around the Sun, I found myself on a ferry crossing Nantucket Sound on my way to Martha's Vineyard, to a celebration organized by my former students and postdocs. It was to be part scientific conference, part "roast," part birthday party, and part implicit recognition that the odds were now great that I was past the apex of my existence. Happily, at sixty I had ever so much more to celebrate and reflect on than I had fifty-nine years earlier.

The same is broadly true for civilizations. We have more to celebrate today than we did two hundred, three hundred, or a thousand years ago. Also, for all the sobering realities facing the average human living in the twenty-first century, very few of us, I imagine, would trade our lot for that of the average human living in any preceding century. On the deck of the ferry that day, I thought of this with regards to the technological advancements we now enjoy. A reason most humans today wouldn't trade places with an average human living, say, one hundred years ago is they wouldn't want to trade in their contemporary gadgets—their Teslas or their iPhone 14s, say—for what then was generally available—a Model T Ford or a candlestick phone, say.

We needn't look a hundred years back in time, though, to grasp the point. The boat I was on had been in service for years, and was therefore older, slower, and more plodding than any number of boats passing it. Those newer, faster, more specialized craft would all reach the island before me. And in that simple insight is great news for space archaeologists, especially those of us hunting interstellar objects of possible extraterrestrial manufacture.

If the archaeologists' task on Earth is to dig ever deeper to discover preserved artifacts from ancient, less-advanced human civilizations, in space the task will be to sift the surfaces of planets in expectation of discovering the last, best thing an extraterrestrial civilization put in our path.

All the spacecraft launched by humanity in its first century of space exploration travel at similar speeds of tens of kilometers per second. This is, of course, the speed limit set by the tyranny of the rocket equation. But we are on the cusp of ending that tyranny. Imagine our second century in space exploration, during which we will develop faster propulsion methods, like the lightsail technology already in use and under consideration at the Breakthrough Starshot Initiative. The payloads of these crafts will not be burdened by fuel, allowing them, in principle, to reach speeds very near that of light.

Yes, the performance of our past rockets is now not so very impressive. Of the five probes that NASA launched toward interstellar space, only Voyager 1 is confirmed to have passed beyond the heliosphere, the interface between the Solar wind and ambient gas. That trip took just under fifty years. That is inspiring, until you shift the scale. Over the past half century, these five craft have traversed less than a tenth of a percent of the distance to the nearest star. A lightsail craft moving near the speed of light, however, could surpass their distance a mere day after its launch. It would be like the speedboat that sped past my ferry, allowing its passengers to disembark well before I did. A consequence by corollary is that as humanity's more technologically advanced crafts are launched, as they travel through the vast expanse of cosmic space and time, our most sophisticated craft will outpace those from our past.

This means that an extraterrestrial civilization near another star is more likely to first encounter our most advanced propulsion systems before much later taking note of the slower-moving spacecraft of our technological infancy. This is especially true if two other predictable variables are accounted for.

One is material strength. It's not just the means of propulsion that improve over time, but the materials we use, informed by our ever-greater knowledge of the interstellar medium. If a mission calls for a craft that is expected to persevere over an interstellar journey, it will be built with the best components and according to optimal designs available at the moment of its manufacture. For example, in April 2022, NASA announced the creation of Alloy GRX-810, which is capable of withstanding temperatures over 2,000 degrees Fahrenheit and developed using a 3D printing process. Constructed to resist fracturing, to retain some flexibility, and to endure extreme conditions, it is the sort of material a civilization would construct if it wanted to send a meteor-carried message.

The second variable we need to account for is profligacy. Costs for spacecraft are highest at the dawn of a space-traveling civilization's existence. These costs include not just the expense of manufacturing a craft but the employment of generations of technologically adept engineers and scientists. Humanity is well aware of the fact that global crises and commercial interests competed for those funds and expertise during the early decades of our space exploration. One result is we can now only boast of five slow Earth-launched craft headed into interstellar space. But what if these costs were substantially reduced and the difficulty encountered in sending out a million lightsail probes was not materially different from sending out one? A predictable result would be the dandelion effect. At such low costs, the temptation to spread many million such probes would be as easily met as a child encountering a field of reproducing dandelions.

While we are envisioning a calming white field of seed-bearing dandelions is as good a moment as any to slay one particular monster that has stalked humanity's early consideration of extraterrestrial civilizations: marauders.

For many decades, our governments and entertainers have fed concerns about the possibility of marauding aliens intent on our planet's plunder and destruction. Presumably based on lessons from humanity's past, the fear of "space Vikings," whose existence is predicated on pillage, is raised to discourage humanity's efforts to explore their cosmic neighborhood, and consequently make our existence even more visible to any potential neighbors. I largely discount this concern. It is because I reason my way to the belief that an interstellar civilization must end up putting their best foot forward, and not just technologically. Culturally, too. A civilization capable of exploring space will display several traits: diminishing ignorance, rising technological prowess, and a shared wisdom about the astrophysical puzzles that make up sentient intelligence's common inheritance. Foremost of these are the nature and extent of matter and the expanding expanse of spacetime.

Another reason few of us would want to live an average person's life during the eras of Leif Erikson, the Spanish Conquistadors, or European imperialism is that violent, extractive exploitation was civilization's then default. By the end of the twentieth century, however, an ever-greater percent of humanity has realized that global trade is more efficient at improving collective well-being than destructive looting. And while human history has suffered eras when civilization receded—consider, for example, Europe's dreadful fourteenth century, marked by plague, war, banditry, and religious schism—it is generally true that if a temporary lottery was enforced and all of us were to be projected back in time, each of us would hope to be among those transported back the fewest years possible.

As was poignantly pointed out by Thornton Wilder in his 1938 play *Our Town*, hesitancy at being randomly cast back even just into our own recent past is well founded. Which is in part why I felt the tug of impatience as faster boats passed the ferry that was conveying me to my birthday celebration. A life or a civilization that is organized to think in interstellar terms and at interstellar scale is resolutely, inescapably future facing. Impatience arises not from putting greater distance between us and our past but from the desire to obtain faster the benefits of a beckoning future, which is in part why I am specifically impatient to learn to a certainty that humanity is not alone in the Universe.

It needs to be allowed, however, that there is a defensible case that life is rare in the Universe, that technological civilizations are rarer, and that humanity is the only one. Just concerning ourselves with the second law of thermodynamics, random combinatorics, and the role of multiple feedback loops with different timescales, a sound scientific and statistical case can be mounted that we are alone. Accept that premise and it leads to one consideration. Zero out a god, first-mover, or some pre–Big Bang sentient's original intention, and humanity's intelligent technological civilization is a random but statistically extraordinary fact. We obtain the status of a rarity. That human civilization, with its 2020 scientific and technical prowess, was realized at all is akin to a Universe-sized population of chimpanzees banging away at the keys of the typewriters and eventually composing *The Tempest*.

This is unsatisfying on scientific grounds. After all, an Earth-sized population of chimpanzees eventually reached a point where only one of them needed quill and ink and a few decades to write the entire Shakespearean canon. And while William Shakespeare completed that task all on his own, the rest of human civilization, all arising out of that same population of earlier primates, produced the entire scientific, artistic, intellectual, and technological content comprising our civilization.

From an interstellar viewpoint, the amount of boasting human civilization is permitted doesn't matter. So long as it is bound to our host planet and star, our civilization's future prospects remain unchanged.

If we're the only civilization on the cosmic ladder of civilizations, the rung to which we cling remains the same, though our prospects become less hopeful. Our predicament is, if human civilization doesn't learn to replicate the circumstances that can sustain it elsewhere, indeed everywhere, throughout the Universe, it will end. Accomplishing that entirely on our own seems less likely than accomplishing it with assistance from a more advanced sentient intelligence's technology. If we're alone, our odds of success go down, but otherwise nothing has changed.

Alone or not, our civilization must set itself the task of ensuring that terrestrial life, and human civilization, endure if they are to endure at all. This too is fuel for impatience.

With age comes knowledge. This is indisputable. This truism can be quantified in a great many ways, but on that particular ferry ride one leapt out at me. Among those gathered to observe my birthday would be most of the members of the research groups with which I had collaborated and that had summarized their scientific insights in nearly a thousand publications issued over a (slight) majority of the decades I have lived.

More disputable is the claim that age brings wisdom. The difference between knowledge and wisdom is well caught in the following aphorism attributed to Einstein: "Wisdom is not a product of schooling but of the lifelong attempt to acquire it." What is certain is that civilizations cannot advance without applying both knowledge and wisdom. Indeed, one way to calculate the worth of a civilization is through a peculiar means of double-book accounting: the ledger of knowledge needs to equal the ledger of wisdom.

Committing ourselves to the wise application of accumulating knowledge is humanity's challenge. If we can accomplish that with ever-increasing consistency, humanity will ascend the ladder of civilizations, perhaps aided by the catalyst of an extraterrestrial artifact. If we fail to act on that understanding, the odds that humanity fails, regardless of whether we discover an extraterrestrial artifact, increase.

I do not believe wisdom is teleological, something civilizations are destined to enjoy with an increase of knowledge. I do believe that human civilization, despite its troubled and troubling history, has grown wiser over centuries, sometimes rapidly and sometimes far more incrementally. The future of humanity, and all its searches for knowledge, theoretical and practical, depends on whether we are now entering a period of rapid acquisition of wisdom or not.

On the hopeful side of the accounting of knowledge and wisdom is that indisputable fact that we possess technology, from lightsail propulsion to quantum computing and artificial intelligence, that hasn't yet seen its wisest applications. Our civilization enjoys the opportunity to commence a period of rapidly accumulated wisdom, balancing our books. The challenges we confront, however, are manifold. Most of these are visible in our news feeds—wars, climate change, pandemics, social instabilities, technological hazards, and violence, discriminate and indiscriminate. Ignorance of our challenges is rare. After all, a large percentage of my fellow passengers on the ferry were busy studying their screens. Few of us are unaware of what humanity is up against because the challenges arising from new technologies and local and global events are, courtesy of the World Wide Web, witnessed by an ever-increasing majority of humans. What is to be done about these common challenges is itself a challenge. A life spent in science allows me to simplify the task down to two words: defeat ignorance. I believe astronomy offers one underappreciated insight to help us accomplish this task: perspective.

Terrestrial concerns are fundamentally different from interstellar ones. Recall the memorable Blue Marble photograph taken in 1972 by the crew of Apollo 17. All of life as we know it and civilization as we've managed it depends on one planet in an expanse of black space. The interstellar vocabulary we are now experimenting with, the interstellar frame for considering our prospects we are now learning to embrace, points toward understanding our challenges from the perspective of eventually untethering life and civilization from our host planet.

We must intend the preserving of terrestrial life and civilization elsewhere in the Universe, we must exhaust opportunities to discover evidence of an extraterrestrial technological artifact, we must put ever more of our collective focus on the task of ascending the cosmic ladder of civilizations. It is from this perspective that three aspects of terrestrial life present us with interstellar puzzles. Teasing out their implications, by metrics of knowledge and of wisdom, I believe will go far toward establishing whether or not humanity adapts in ways that make our interstellar success more probable.

First, the Universe's vast but not infinite generosity to humans. Your body, its bones and organs, its three-pound (on average) brain, and everything these interact with are all made of heavy elements that were fused in the cores of massive stars. Those stars exploded, enriching the interstellar medium, out of which the Solar system formed, inside of which the Earth condensed, and on which life emerged, was nourished, and as a result humans arose to chronicle our temporary existence. Viewed this way, we are merely passengers on a space-born cabin called Earth, forging a civilization by rummaging about in borrowed luggage, all the while heading to an unknown destination. From the ground you walk on, the air you breathe, and every material object you can touch, all of it has been constructed of the materials the Universe gifted us. On Earth, the species *Homo sapiens* is collectively the greatest beneficiary of that generous inheritance. We can, however, claim a few things as uniquely our own. These are ignorance, knowledge, and applied wisdom. How to mix our own contributions with our material inheritance is our interstellar civilization's immediate mystery to solve.

That brings me to my second astronomical puzzle, the span of an average human life. It is limited to a period that is at least 100 million times shorter than the age of the Universe. Our lifespan, which on average amounts to 72.5 years (globally women average 75 years, men 70 years, though this number is apt to drop post-COVID pandemic), allows us to individually experience only a brief snapshot of cosmic history. To offset our wink of a lifespan, we have knowledge. Science, the tools of science, and the iron law of evidence expand our individual perspectives regarding the big picture.

It is possible that there are fixed limits to human knowledge. Perhaps the example most often postulated is consciousness. What it is and how it arises could be beyond human ken. A less likely limit is proving (or disproving) dark matter, or arriving at a unified theory of physics. Whether these or other mysteries of existence become accounted for, with knowledge and wisdom zeroing out a line item of ignorance, will take generations to discover. And that is time best spent when we make correcting our common ignorance the work of common effort.

Humanity has come a long way toward making the scientific exploration of our shared existence a common enterprise. A few hundred years ago, scientists were separated by oceans considered vast and woefully inadequate means of communication, and pitted against one another due to explicit racial and gender biases, explicit nationalist jealousies, and competing strains of religious fundamentalism that privileged ignorance over knowledge. Over centuries, and more rapidly over the most recent decades, all of these impediments have receded in importance. (Receded, not disappeared, of course.) Slowing our advance is that for thousands of years, and down to the present, we have individually chased knowledge in pursuit of consolation prizes in the forms of honors, awards, pecuniary prizes, and—today—"Likes" on social media. For centuries, and down to the present, some guardians of data have put narrower interests ahead of common ones.

To take just one example, consider the United States government.

We now know that for decades it collected data, anecdotal and better-than-anecdotal, about unidentified aerial phenomena. How much better remains an open question. An unknown extent of the data the government possesses is kept secret. From personal experience, I know that the United States government holds data capable of confirming, or not, interstellar meteors. The United States is understandably focused on concerns involving adversarial nations. But when data of extraterrestrial or interstellar nature are not shared openly, scientists can feel as I did during the years-long wait for the hypothesized interstellar origin of IM1 to be confirmed by data held in secret since 2008. It is a feeling akin to the child in the Hans Christian Andersen story who alone speaks to the fact that the emperor has no clothes. In my case, it was not just that the United States government was acting as if it had no relevant data. I knew to a certainty it controlled a closet full of data, though how big a closet and how overstuffed remains unknown.

Withheld data inflict manifold harms. A theory the hidden data would disprove is allowed to take hold. Hypotheses the hidden data would encourage remain unthought, or unfairly criticized. Because encounters with UAP were kept under a veil of secrecy, it became easier to stigmatize the few pilots brave enough to speak out. The United States government admitted as much in the June 25, 2021, report of the Office of the Director of National Intelligence. In it the government acknowledged that UAP data are rarely discussed openly because of "sociocultural stigmas" and that "reputational risk may keep many observers silent, complicating scientific pursuit of the topic." Sometimes, this means fear of stigma causes real but unknown risks to go unaddressed. Lieutenant Graves reported that after two pilots passed to either side of an apparent dark box in a translucent bubble, they cut their flight short and returned to base. They did this not out of fear of encountering aliens, but out of fear of a midair collision with unmarked, poorly visible objects in their restricted airspace. Shutting down the discussion of an encountered risk is to pretend such risks don't exist. If humanity is to avoid the experience of trying to knit the emperor's clothes while he walks, increasingly humanity's default should be shared, not secret, scientific data.

There is lots of evidence for the efficacy of this approach.

In the late 1960s, the United States launched the Vela satellites to detect high-energy radiation, also known as gamma rays, emitted by nuclear weapons tested in space. The government was concerned that the Soviet Union might attempt to conduct secret nuclear tests after signing the Partial Nuclear Test Ban Treaty in 1963. On July 2, 1967, the Vela 3 and 4 satellites detected a flash of gamma radiation unlike that expected from any known nuclear weapon. Alarm bells sounded in Washington. Uncertain of the implications, a research team at the Los Alamos National Laboratory, led by Ray Klebesadel, filed the data, which were never held in secret, for further analysis.

It would have been natural for the United States government to initially regard the unexpected gamma-ray flashes as a matter of national security. But as additional Vela satellites were launched with better instruments, the Los Alamos team continued to identify gamma-ray bursts. By analyzing the different arrival times of the bursts to different satellites, the team was able to rule out a terrestrial or even Solar system origin. The data were not classified, and the discovery was openly published in 1973 as an article in *The Astrophysical Journal* entitled "Observations of Gamma-Ray Bursts of Cosmic Origin." It took a couple of additional decades before the cosmological distance scale of the bursts was established through X-ray localization by the Italian satellite BeppoSAX. As a result, humanity scientifically identified the first gamma-ray burst. While these explosions originate at great remove from Earth and last milliseconds to a few hours, they represent one more potential threat to life on Earth. Knowledge of gamma-ray bursts is far preferable to ignorance.

A truism for humanity is likely a truism for all biological life in the cosmos. Evolution, which selects species to survive long enough to persist cross-generationally, is a backward-looking mechanism. Humans randomly selected to better survive the last plague may, or may not, be randomly selected to survive a future one. Of all life on Earth, only humans possess the skill set that makes life-as-we-know-it existing past the due date of our Solar system possible. This is also why it will be only at the scale of civilizations that the collective efforts of an advanced technological species might leave evidence of its existence in a way commensurate to the scale of the Universe. Which introduces us to my third astronomical puzzle, and perhaps the most disconcerting: the temporal scale of space.

The trajectory of my individual life, from my birth in Israel in February 1962 to, as of the time I write this paragraph, a scientific conference by way of a birthday party on Martha's Vineyard in

February 2022, seems overstuffed with unique particulars. But any sense of individual moment and consequence erodes to insignificance when measured against the vastness and age of space.

We are by turns awed, horrified, amused, and bemused by the moments we experience in our, on average, seventy-five years. We are impressed by the chronicled accomplishments of our species. Yet biological conditions on Earth are likely replicated in nearly sextillion (10 to the power of 21) habitable exoplanets within the observable volume of the Universe. We rarely allow our inescapably obvious insignificance within the Universe to intrude on our habit of terrestrial navel-gazing. And it is this habit that helps explain why most of the scientific community, aware of the Universe's cornucopia of exoplanets, still regards the claim that other intelligent civilizations existed over the past 13.8 billion years since the Big Bang as extraordinary.

One way to reconcile the vast scale of the cosmos with our limited perspective is to admit as a starting point that we are not extraordinary. The news provides a daily reminder of how much room there is for human civilization's improvement. The cosmos cares so little about us that it didn't bother to provide us with a manual for our existence. It no doubt extended the same lack of courtesy to every other sentient life in the Universe. But therein lies hope, a human lesson of common knowledge and wisdom arrayed against the weight of common ignorance applied to the Universe. Might we find other passengers who had been or even have been around longer than we? Recall that with age comes knowledge and, often, wisdom. And with wisdom's expanse, what was once held in secret by some becomes what is held in common by all. Our best hope for success over failure among the civilizations of the Universe is to bet on our unextraordinary existence and that universal ignorance gives way to pursuit of common knowledge, and perhaps shared wisdom.

There is one clear difference between the scientific search for confirming evidence of dark matter and the scientific search for confirming evidence of an extraterrestrial civilization. There is no evidence that dark matter is conscious, and abundant evidence that its existence, its particles, await without interest our discovering them. To all matter, dark or otherwise, whether humans discover its fundamental properties is of no importance simply because matter is not sentient.

Not so for any extraterrestrial civilization. By definition, extraterrestrial civilizations were, or are, collections of sentient, decision-making life, and if they possess substantially more advanced technology than humans their options are likewise substantially more advanced. Among those options would be to deliberately conceal their existence from us. Why is best understood through a simple thought experiment: if humanity had been given the choice to either suffer the global COVID pandemic or not, every sane human, with knowledge of the Spanish flu, common influenza, or a bad cold, would have wisely opted for no pandemic. The COVID-19 virus, no more sentient than matter, didn't give us the choice, for it exercised none itself. That will not hold for another extraterrestrial technological civilization. Proof of another civilization could well depend on its willingness to disclose itself to us.

It may well be that our still-sentient cosmic neighbors in the Milky Way galaxy chose to swipe left while monitoring our daily news on the universal civilization-scale dating app. Or, they might take a more maternalistic position, acting the Good Enough Civilization and hoping to nurture our true independent maturity. Perhaps the Universe's civilizations act similarly to the psychologist D. W. Winnicott's "Good Enough Mother," who empathetically allows a child to encounter, and work through, frustration and errors on the way to maturity. Or perhaps other civilizations are neither self-protecting nor empathic. Instead, they are as disinterested in humanity as humans are of most animal life on Earth. If this is the case, we can hope they are as profligate with their trash, for it would mean we could be as richly rewarded as raccoons and flies are by human carelessness with garbage. Regardless of its makers' intent, I would be thrilled to hold an advanced civilization gadget-as-garbage, and no matter if that civilization were extant or not.

FROM VIRTUAL TO ACTUAL REALITY

Our knowledge about relativity and quantum mechanics was acquired about sixty years before I was born. This means my life over the past sixty years has spanned half of modern science. I sometimes muse about the technological and scientific advances realized in my lifespan, from streaming services to the James Webb Space Telescope, and eagerly anticipate the achievements that await us. Contemplating the possibility of extraterrestrial civilizations, however, only hones my impatience. My wish for the remaining decades of my life is that I will have the opportunity to learn from a scientific knowledge base that was acquired over a period of time that is 100 million times longer than humanity's. Knowing that someone else in our cosmic neighborhood did better than us will ease my mind. It will replace impatience, a useful definition of helplessness, with knowledge, a useful definition of the scientific method.

Humanity is increasingly as impatient as I am for answers to questions like "Are we alone?," "Who else is out there?," and "Might they be helpful?" Within the next few years, the Galileo Project's search for indisputable evidence of an object that is not natural or human-made will make substantial strides forward. The Project just might provide a simple answer to Fermi's paradox. The answer to "Where is everybody?" will no longer be a shrug, but rather a confident "There, there, and there!" The instruments developed and deployed by the Galileo Project represent a brand-new observatory design with unprecedented capabilities. As ever more of them are built, perfected, and deployed across the globe, they can become our best way to find out whether we have curious neighbors. If extraterrestrial technological dandelions are blown our way, our new instruments and software will help discover them. And finally, we will continue to refine the expedition parameters that will direct the effort to retrieve fragments from the first interstellar meteor from the ocean floor near Papua New Guinea.

The latter, of course, is the most tantalizing part of our project. No distant prospect, but rather a discovery potentially just months away, if we elect to expend the effort.

Pause over that. In Chapter 5, I explored the tyranny of the rocket equation, or the need for an exponentially increasing amount of fuel to realize modest, arithmetic increases in speed. That tyranny speaks to craft that humans have sent into space. But everything in the Universe is moving and everything that moves does so because of forces acting on them. Some set of forces sent IM1 moving at the speed at which it hit Earth. There are certainly a number of naturally occurring forces that would account for its speed, and assumptions about statistical variables allow for a random encounter with Earth. For example, this first interstellar meteor and unusually tough rock could have been ejected from the vicinity of an unusually fast-moving star and by chance was set on a course that guaranteed its encounter with our planet. Another explanation would be that it is a technological relic from another civilization directed here by intent or not.

The only way to find out is by scooping up its fragments and studying their composition in the laboratory. This is the work of curiosity-driven scientists eager for new knowledge. It is also the work of elements of our civilization that wisely prioritize this expedition over others. Concretely, it requires that we triangulate the impact location (done), identify a boat and the machinery needed for scooping the ocean floor (done), raise a couple million dollars to fund the expedition (done), and undertake as many expeditions as are necessary to see what is resting on Earth's surface.

Which leads me to a new way to frame an answer to two questions: "What is the likelihood humanity will succeed or fail to discover an extraterrestrial artifact?" and "What might be the consequences?" I recommend we approach both questions by first considering another: "How would our civilization compare to other civilizations that have existed since the Big Bang?"

First, a few words about how I'm making these judgments. While there will be unknown differences across civilizations, there will be universal commonalities. One will be knowledge of our shared Universe. Another will be comparative technological competence based on that knowledge. Finally, wisdom. This will be captured by how a civilization translates its experience and understanding of its circumstances into sound judgments about confronting the Universe's risks and opportunities. How to measure these qualities—knowledge, technological competence, and wisdom—requires some consideration.

I believe civilizations that prioritize science should be valued higher than those that don't. The greater the fraction of its gross domestic product that is used on scientific research and exploration, the higher it ranks. This is for the simple reason that accumulating evidence-based knowledge allows a civilization to cope better with the challenges it faces, whether those challenges be disease, environmental catastrophe, or space travel. It is better for a civilization to attend to the well-being of its planet and the life the planet supports. It is better for a civilization to dedicate more of its productive wealth and capacity toward the increase of knowledge and the decrease of ignorance. Not only will such civilizations have better prospects for longevity. They will also leave a discoverable technological legacy.

Scientific knowledge is key to exponential growth in developing new technologies. But as I've already noted, knowledge without wisdom confronts limits, usually captured in an overabundance of technologies uniquely adapted to destruction or escapism.

To track the reach of wisdom in a civilization, I propose two other key considerations. One is the spirit of exploration of our shared reality. And I emphasize the word *reality*. A civilization that is fully immersed in wearing goggles and enjoying virtual realities rather than venturing into the vast space that lies beyond the surface of the planet it was born on will be vulnerable to single-point destruction by a catastrophic event. The second key factor is a focus on constructive rather than destructive goals. These two factors are entwined. A civilization riven by factions focused on global acquisitions obtained by threatened and actual destruction is one that has pursued ignorance to arrive at a habit of cruelty that predicts extinction.

It is helpful, though bracing, to think of the Universe's civilizations as each playing a game defined by continuance—you win—and extinction—you lose. While the talents of a civilization are unique to it, the rules as captured in the physical laws are the same for all. What makes the metaphor bracing is there are no do-overs.

A curious and constructive civilization that is focused on increasing its scientific knowledge base to better allow its technological exploration and colonization of interstellar space would be the most promising participant in this galactic contest. Such a civilization is also more likely to be drafted by others seeking support, or perhaps just like-minded company. Less capable players, with correspondingly weaker skill sets, are likely to spend most of their time on the bench and never get invited to play as members of the cosmic dream team. Of course, the lowest-ranking civilizations might be spared that embarrassment. The lowest-ranking civilizations would not even know there was a bench from which the ambitious, hopeful, and determined could watch the highest-ranked participants perform.

Among interstellar species, the guiding principle is likely to be survival of the wisest. This sounds similar to the distorted claim, survival of the fittest, lifted from evolutionary biology and Charles Darwin's discovery of natural selection while exploring the Galapagos Islands in 1835. In space, I suspect, Darwin's momentous terrestrial accomplishment, captured in his *On the Origin of Species*, will be of little interest.

There is little need for Darwinian natural selection across the ocean of interstellar space since life as we know it would have formed in galactic islands of isolated planetary systems. And transfer of life among planetary systems, commonly labeled panspermia, is unlikely to occur by natural means because of the long journey time, lasting millions of years, the small probability of delivery, and the hazards posed to life in interstellar rocks by cosmic-ray sterilization. Additionally, natural selection doesn't apply to asteroids and comets any more than it does to pebbles and stones. Among sentient intelligence in interstellar space, what is most likely to determine survival are choices made by the civilizations. Civilizations that survive will be those that first select to avoid self-destruction long enough to venture into space and that then select to manufacture the technology needed to survive without dependence on host planet and host star. Among these civilizations, the most advanced, space-venturing cultures will be those that self-select for knowledge, wisdom, resource commitment, and a creative spirit of exploration. These civilizations will be the ones that most thoroughly give themselves over to the task of excelling at scientific and technological selection.

I see cause for optimism. Humans have, on balance, technologically selected more often the virtues of health, comfort, curiosity, and joy than the dangers of mutual extinction. Yes, we manufacture weapons of destruction, but we manufacture many more phones, computers, and bandages. Yes, externalities to our energy creation do threaten life on the planet, but as much as we now wish to mitigate against the pollutants warming our planet, few wish we could transport all humans to a preindustrial existence. Rather, today's intent must be to exercise greater knowledge and wisdom than our ancestors did. I remain optimistic, for example, that in the long run we are more likely to use nuclear energy for propulsion in interstellar space than for mutual annihilation on Earth.

Terrestrially, we are ever more technologically selected for survival. This is, perhaps, preparatory work to becoming adept in space, where technological selection rules. It also means that, from our vantage point, our early attempts to reach interstellar destinations are less significant than our future efforts. It is reassuring that our most advanced ships will bring the flag of sophistication first to the attention of extraterrestrials. As our science and technology progress, our crafts could reach farther destinations, always with humanity's best foot forward.

There is urgency. The race through interstellar space may be more consequential than we imagine. Technological selection is competitive for a good reason. The surviving products of the most advanced technological cultures since the Big Bang may seed the most prominent forms of life across our cosmic neighborhood. A single intelligent culture could have spread the seeds of life throughout the Milky Way galaxy. Discovering that we might be one of these seeds would have a profound impact on the meaning we assign to our life. Given the late birth of the Sun and our Solar system, we come to this practice of civilization later than others. And the ground we need to make up is increasing, literally.

A trillion years from now, the Universe is likely to go dark. Any astronomers still then active will have only their own galaxy to examine, a result of the accelerating expansion of the Universe rendering all extragalactic light invisible. Such astronomers would be at an insuperable disadvantage. Consider that our cosmological understanding of everything from supernovae and cosmic microwave background radiation is a result of an observable Universe to observe.

Even to this problem, however, technology might offer a solution. When the accelerated expansion of the Universe was discovered more than two decades ago, I published a paper about its gloomy implications for the long-term future of astronomy and humanity. Once the Universe ages by a factor of ten, we will be left in the darkness of empty space without any external galaxy visible beyond our own. The astrophysicist Freeman Dyson emailed me to suggest a possible means to avoid this otherwise guaranteed loneliness. Dyson hypothesized a civilization capable of harnessing the gravitational energy of a group of galaxies that would be sufficient to keep at least a local group of them tethered despite the accelerated cosmological expansion.

This sounds like a fantastical consideration, the sort of exchange of ideas two astrophysicists might have if they had too much time on their hands. Of what practical use, it could be asked, is even the theoretical consideration of how to bind galaxies together to withstand cosmological expansion? What astrophysics teaches, though, is that at some point this is no theoretical question. Accelerated cosmological expansion is established fact; the expansion speeds of visible distant galaxies are growing ever faster and will eventually exceed the speed of light; if whether for aesthetic reasons or colonization, binding some of them to each other is valuable, doing so starts with theory. If humanity decides longevity is preferable over extinction and that knowledge impowers in ways ignorance does not, and after obtaining a civilized level at which we are no longer actively working to cook our planet, then our interest in persisting off of Earth will become obvious.

The answer to the question whether or not humanity will succeed in its attempt to ascend the ladder of civilizations, perhaps with the assistance lent by the relic of a far more advanced technological civilization, turns eventually on only this: Will humanity decide that humanity is worth the effort? Deciding in the affirmative will be a self-fulfilling prophecy. Some, ideally most, humans selecting to take up the work of helping ensure terrestrial life and human civilization endure is the necessary precondition. Perhaps it will prove something more than that, too. Perhaps it will be equivalent to an invitation to curiosity. After all, as any baseball or basketball hopeful knows, to attract the interest of a professional scout you had better first make obvious your diligent practice, effort, and attained skills. Only then are you apt to be considered for a spot that, after further effort to distinguish yourself, might warrant someone printing a trading card with your likeness on it. A given, however, is that if you don't make your talents visible, you will go ignored, perhaps by accident, or, worse, with intent.

7

SURVIVAL OF THE OPTIMISTS

THE PENTAGON REPORT ON the existence of UAP posed many explicit questions. First, and most obviously, what, precisely, is out there? And what are the necessary scientific steps that now need to be taken to start to provide humanity with answers? There were also clear practical questions concerning risks. How concerned should airplane pilots and commercial travelers be? In addition, there was an interest in a broader threat assessment. Do America's adversaries, terrestrial or not, have capabilities that the country needs to prepare countermeasures for? Finally, there were the obvious, implied questions. Why was information concerning UAP treated as classified, and what remains classified?

One of the most profound questions that UAP raise, however, goes unasked in the report: Why do we think an extraterrestrial civilization would pay us any attention at all? Answers to that question have practical implications, such as where we should place Galileo Project UAP observatories. If we had reasons to believe UAP are more attracted to our oceans than to our cities, we would place our observatories accordingly. If we had reasons to believe UAP are more attracted to our military conflicts rather than to our seats of governments, we would calibrate our sensors differently. More pressing is the understanding that answers to that last question would also have philosophical consequences. What are the characteristics of human civilization that we think are most deserving of interest? For what reasons do we hope a more advanced civilization would pay us a visit?

To appreciate these questions from the extraterrestrials' vantage point, consider some possible destinations for your next vacation.

If you wished to, you could pick a month in the summer to visit the Danakil Depression in Ethiopia. There you will find lakes of bubbling mud and lava that burns blue, all courtesy of plate tectonic drift and escaping sulfuric gases that will burn your lungs and cause your skin to peel. The depression is due to the Earth's crust widening at its cracks, literally. Not to your taste? How about a winter month atop Mount Washington in New Hampshire? There, far from any typhoon or hurricane, you'll likely encounter winds of 200-plus miles per hour. And the normal daily minimum temperature at the summit in February is negative 3 degrees Fahrenheit, with a record low of negative 46.

It would, of course, be entirely reasonable for you not to visit well-documented toxic, inhospitable, or otherwise dangerous places. And that is the purpose of this very brief travel brochure of risk. The hard truth is that humanity may not have to worry about encounters with extant extraterrestrial civilizations, and their elusive UAP, because there is a good chance that they are making just such a choice by steering clear of us.

Can we really blame them? Even some humans are sincerely uncertain whether our civilization is suitable for future human life. The journalist and cultural critic Ezra Klein says the following is the question he is asked more than any other: Given climate change and the deteriorating environment, should worried adults opt not to have children? In an opinion piece for *The New York Times,* Klein gave the question thoughtful consideration, and declared himself a blindfolded optimist. The climate, and with it the world, confronts a litany of troubles. And yet: "I can't tell you whether" having children "is the right choice for you, but no climate model can, either."

Because I am a wide-eyed optimist, I believe the opposite. The Earth's demise is an inevitability, whether it is a few years off, due to an act of egotistical human hubris or a massive meteor not yet identified, or a billion years off as a result of the Sun's expansion. We may balk at exposing an innocent child to such horrors—kids, lacking experience, are bruised so easily. Yet humanity's only hope and our civilization's one chance at persisting rests on the knowledge, optimism, and habits of thought and behavior of some of our children.

JUNIOR MOMENTS

We adults likely have learned how to avoid lingering on the scary forecasts of collapse, just as we've learned to minimize risk in day-to-day life. We know better than to take a trip to the Danakil Depression. That we do is entirely owing to others preferring risks to ignorance. After all, the only reason we know anything about the top of Mount Washington is because intrepid others have been there, gathering data.

Adults who wish to protect themselves from harsh realities can overcorrect and immure themselves from the truth, whether by real or virtual walls. They can spend a lifetime insisting on wrong notions and building psychological blinders that isolate them from evidence that would falsify these notions. We should follow our kids' examples and plunge into the unknown—it may lead to a bruised shin or ego, but such bruising is the only way to learn something new. Recall that in Hans Christian Andersen's parable about the naked emperor, the person who found the absurdity of the world impossible to ignore was a child.

This pattern, failing to summon the courage to confront an absurd but perhaps beloved notion, is not only apparent in politics but also in the frontiers of science. Except, there, through technology, we can use the pattern to our advantage. It was with this in mind that I brought up the use of machine learning in identifying anomalous objects in the sky during a Galileo Project team meeting. Our artificial intelligence algorithms will initially have little experience and will make mistakes, just like kids. But as the algorithms analyze more data, their library of possible objects will eventually grow to the extent needed to classify the vast majority of all objects detected by our telescope systems. At that level of experience, it would be straightforward for amateur computer algorithms to identify outliers. Heck, at that level of cataloged experience even a child could do it!

The main advantage of computer algorithms relative to humans is that they do not become more worried about what others might think of them as they mature. Where humans may balk at drawing conclusions that if uttered aloud might sound absurd to powers-that-be, an egoless algorithm will have no such reservations. For us humans, age correlates with increased experience, which means that being older is often assumed to confer authority. But the nature of discovery involves new knowledge, and the most valuable discovery is invariably the one that the reigning authorities on the subject least expect. Human history across all domains overflows with examples of experts suppressing innovation because it threatens their pride, their professional reputations, and their self-esteem.

The startling insight of the young rebel becomes his later, gatekeeping self's most prized accomplishment at which more younger rebels are aiming slings, arrows, and data.

During my career, which overlaps with a bit more than a third of the one hundred years of modern terrestrial science, I have witnessed how some of the most exciting current frontiers of knowledge were ridiculed in their infancy by established experts. Examples include studies of the early Universe, extrasolar planets, and gravitational wave astrophysics. After evidence mounted to a level that could not be ignored, these disciplines went through phase transitions in which the rebels who discovered that evidence became the expert gatekeepers of the new, brilliant insight.

That trajectory, rebel to gatekeeper, is seemingly entrenched, not as a feature of science but of human ego. It is not so much a flaw in evolutionary biology, but more probably endemic to our civilization's governing cultural habits. Recall again that the human body is built to survive, on average, around 72.5 years. Many more of us are living out the full term of these seven or so decades than was true in the recent past, and a rising minority of us are living even longer, but the data suggest that the outer limits of our longevity have been more or less set for a very long time. Tellingly, humans routinely live a few decades past optimum fecundity, nature having selected for slow growth—slow to mature, slow to expire—to, perhaps, allow for cross-generational encouragement. Our genes may be selfish, but conscious humans needn't be, especially with their later years. Too often, however, we work within a culture that fails to take full advantage of this fact. Too often, our productive lives follow a pattern of existing long enough to find a flaw in older generations' efforts, assuming the authority to advocate for the paradigm-shifting corrective, then spending the next decade or two defending that corrective, with breaks for golf and committees. The extra years gifted us by evolution have, in our civilized hands, too often given rise to a tug-of-war between accomplishment and comeuppance.

One answer to why an extraterrestrial civilization might be paying attention to humanity is for the sake of our children. To an extraterrestrial civilization that has survived, evolved, and improved over tens of thousands, even millions, of years, humanity's past is apt to be of no more interest to them than our present. Our future, with its latent possibilities to be influenced by the conduct of our civilization, is another matter. Human civilization, after all, has attempted to construct cultural pipelines for the discovery and advancement of talent. Harvard University represents one apex of this inefficient, flawed, prejudicial, but still aspirational process. Our globe is dotted with them.

UAP might be usefully considered cosmic civilizations' highly selective college recruiters. They would conceal themselves but observe us in hopes of capturing, after all those millennia and light-years, the promising sequence of what are best thought of as humanity's junior moments that signals a civilization grasping the next rung up the cosmic ladder.

When old people forget names, they apologize for having suffered a senior moment. I think this has its corollary when the adults in the room dismiss innovation by young people. The experience of youthful enthusiasms and insights that are undercut, perhaps outright shot down by the older authorities in the room, only to later triumph over calcified wisdom; this deserves a label of its own. Let's call it a junior moment. In my experience, professional life is worth living for these junior moments. They hold the greater promise for a reboot to our thinking, and these opportunities are often not appreciated by the reigning authorities. Once you lose interest in the thrill of junior moments, you turn professionally into dead wood.

Let me be clear. The transition to dead wood is a matter of choice, not of biological age. Yes, there are currently inescapable bookends. Tragically, for some of the elderly, dementia awaits, and at the endpoint of the timeline death awaits us all. The other age extreme has its pitfalls as well: a two-year-old's solution for everything from consuming food to traversing a room is, we know, suboptimal. But between these poles of existence, the scope of time for junior moments is not fixed. I know of many older scientists who are innovative risk takers, and I know fledgling scientists who walk the beaten paths because of the peace of mind it brings. Choosing to prolong the chain of your junior moments and refusing to assume the status of a gatekeeper are options that remain available, regardless of age.

Scientists would do well to recall the exchange from the movie *The Wild One*: "What are you rebelling against?" a woman asks a young Marlon Brando. "What've you got?" is Brando's reply. For the rebelling scientists, the answer fits, but only because they are rebelling *with* a cause. They are in search of junior moments, the scientific questions freighted with transformative possibilities. Those are the causes, the domains of explorative study, of the truly rebellious scientists, the truly rebellious humans, no matter their age.

The current fact is, we do not know if extraterrestrials or their technology are visiting us, appearing fleetingly as UAP, and if they are we certainly do not know why. Filling those yawning gaps in our knowledge with data is the purpose, of course, of the Galileo Project. Knowledge of our ignorance is the starting point of all science. Acting deliberately to gain knowledge to address that ignorance is the first step of a civilization self-selecting to exist. Among interstellar civilizations, perhaps it is something additional. What would be the most enticing invitation we could extend to interstellar civilizations vastly more evolved than humanity's? Perhaps it is demonstrating the robustness of our rebellious curiosity. Given our ignorance, we must also consider that we are astounding statistical outliers, the only sentient life in the Universe, and our technology the apex of technological achievement across the vastness of space and time. In which case, persisting is entirely up to us, which I believe means it is entirely up to encouraging our optimistic, courageous rebels who are chasing junior moments.

For me, there is a tremendous thrill in encouraging the junior moments that will lead to insights that I will not live to see, but for which I helped lay foundations. I can acknowledge that thrill precisely because it is one I have experienced in real time. If UAP are out there and paying attention, it is these sorts of events, the ones that hint at humanity's capacity for scientific progress, that I hope they monitor. Consider just two.

By the year 2000, astronomers had identified a correlation between the mass of black holes at the centers of galaxies and the luminosity of the spheroid of stars that surrounds them. As I considered this discovery, it occurred to me that any such correlation was probably the result of the black hole saturating at a mass where its powerful energy output expels the gas reservoir that feeds it. The process would be similar to a baby who has consumed all it can and lets his parents know by shoving the remaining food off the table. The growth of the black hole depends on the depth of the gravitational potential well that keeps feeding it, similarly to the depth of a bowl containing the baby's food. With time, a parent learns to read the signs, put down the spoon, and avoid the mess. Similarly, an astrophysicist can with time learn the depth of a gravitational well by gauging the velocity dispersion of stars around the black hole.

At a 2000 conference in Leiden in the Netherlands, I aired this notion publicly. I proposed plotting the correlation between black hole masses and the velocity dispersion of stars in their host spheroid of stars. The idea was quickly dismissed as uninteresting or impractical by the experts in the room. Yet, on returning to Harvard, I attended two lectures by job candidates for the astronomy department, Laura Ferrarese and Karl Gebhardt, each of whom presented a correlation between black hole mass and spheroid luminosity. In separate meetings with them, I suggested that they plot the correlation with velocity dispersion instead. Two months later, I received an email from each of them informing me that they had independently discovered that the correlation with velocity dispersion is tight and that they, along with their research groups, were about to submit papers on the subject. Eureka: a junior moment! Followed almost immediately by a senior one: over the following years, the two teams proceeded to fight fiercely among themselves for the credit of being first to derive the by now well-established correlation between black hole mass and velocity dispersion of stars in their host spheroid.

A couple of years later while on sabbatical, I recognized that imaging the motion of a "lightbulb" just outside the Innermost Stable Circular Orbit in the highly curved spacetime of a black hole could establish a new test of Einstein's theory of gravity. I conjectured that such a lightbulb could be realized naturally through a "hotspot" in an accretion disk of gas, heated through reconnection of magnetic field lines that cross each other, similar to the flares rising off the surface of the Sun. When I suggested this simpleminded idea to a few local experts, they dismissed the notion of a "hotspot" as unrealistic. Given the turbulent dynamics of gas near a black hole, they argued, any hotspot will quickly dissipate through turbulence or be sheared away by the rotating gas. On my return to Harvard, I nevertheless suggested this idea as a research project to my new postdoctoral fellow, Avery Broderick. Over the course of years, Avery and I wrote a series of papers on the observable consequences (such as light curves and polarization maps) of a "hotspot" moving around the largest black hole in the sky, Sagittarius A* (Sgr A*), which resides at the center of the Milky Way galaxy and weighs 4 million Suns.

In 2019, a team of astronomers at the Max Planck Institute for Extraterrestrial Physics in Germany, led by Nobel Laureate Reinhard Genzel and spearheaded by Frank Eisenhauer, announced the observational discovery of hotspots moving in a circle on the sky for three flares near the innermost stable circular orbit of Sgr A*. Their observational data, obtained with the GRAVITY instrument on the Very Large Telescope Interferometer in Chile, confirmed my theoretical idea, born of a junior moment I had fifteen years earlier.

Even within human civilization, these and countless other examples confirm that sentient intelligence pays attention to the advances being made by other sentient intelligence. Of all the reasons I can imagine as to why an extraterrestrial civilization might seek us out, or even nudge along our progress, the most hopeful is our courageous immaturity. That our civilization could be plausibly deemed too inhospitable, too dangerous, perhaps even virus-like, is among the least. Any such visit would amount to the arrival of a cosmic band of Doctors Without Borders, an interstellar Doctors Without Planets. A third possibility, the Zoo Hypothesis, which imagines extraterrestrials, who are technologically able to go and visit exoplanets at will, would find humans sufficiently curious or amusing to stop off and glimpse at. They visit us much as we visit a zoo. I find that hard to believe simply because we're not that interesting. But someday, we may be. And if some UAP are discovered to be extraterrestrial, that could be evidence of our nearing some stage of maturity warranting the interest of an ETC.

What would constitute such evidence? There are some who have speculated that UAP sightings are more numerous around nuclear reactors. I am dubious. Humanity constructed the first nuclear reactor eighty years ago. Yet there was no immediate evidence that this accomplishment attracted attention from extraterrestrial civilizations. This is understandable. Nature routinely manufactures nuclear reactors of much greater scale, in the form of stars, rendering humanity's collective prowess at generating nuclear fission and fusion a cosmic banality. Meteors routinely release as much energy in our atmosphere as our nuclear weapons could. Ditto if we ever prove successful at developing primitive forms of life in our laboratories. Nature fulfilled that challenge quickly on Earth, based on random chemical processes. Nothing to brag about in our cosmic neighborhood.

But what about the creation of a sentient artificial intelligence (AI) system? Could that raise interest from extraterrestrial analogs based on a sense of kinship?

If you have ever taken your dog to a dog park you will have noticed how quickly canine interest in humans is supplanted by canine interest in their fellow canines. The same goes for bringing children to playgrounds. In both instances, the odds are good that you, "the adult in the room," don't feel particularly hurt by the fact that you are an excluded outsider watching the unrestrained joy shared by members of a club to which you do not belong. The opposite, of course, is true when we're denied acceptance to our preferred university, passed over for the perfect job, or rejected by the then love of our dreams. The framework of attraction makes all the difference.

For decades, scientists engaged in the search for extraterrestrial intelligence have acted the presumptive suitor. We have sent out messages in hopes of their being sufficiently alluring to elicit a reply. This illusion of human significance on the cosmic stage is best exemplified by the "Golden Record" on the Voyager spacecraft. Boastful of our past triumphs, in 1977 we put together the equivalent of a mix-tape collection of our greatest accomplishments that included a total of 115 images and ninety minutes of music. The result is a projection of the Zoo Hypothesis to the Universe: our past is worth observing because it is our past. Compare our physical features and joys (one image shows a woman eating an ice cream cone) to yours; compare our most accomplished musicians to yours; compare our grasp of existence to yours. Almost always, this research has anticipated a human biologically evolved sentient life in conversation with another biologically evolved sentient life at about humanity's level of scientific and technological accomplishment. If, however, they vastly exceed us on these fronts, they may regard our boasting as unfounded, our images of human women eating ice cream unimpressive, and our efforts to attract them underwhelming. SETI-like overtures could well be similar to a colleague asking you to view with excitement their vacation photographs after they've visited a place you've been avoiding.

A sentient being capable of roaming the Universe may not be biological at all. Despite the naïve storylines of interstellar travel in science fiction, Darwinian evolution has not selected for the kinds of traits that would allow biological creatures to survive travel between stars. This is almost certain to be true of any creature optimized to persist within the confines of a home planet. The interstellar trip must span many generations since even the speed of light requires tens of thousands of years to travel between stars throughout the Milky Way disk and ten times longer to cross its halo. It is far more probable that a sentient AI spacecraft would undertake that journey. In which case, an AI system that we produce might have a more intimate connection to an extraterrestrial AI system than to us. If some UAP represent extraterrestrial AI systems, perhaps their appearance would be explained by humanity's nearness to developing its own sentient AI.

In the summer of 2022, it was widely reported that the Google engineer Blake Lemoine, on opening his laptop to the restricted interface for LaMDA, Google's artificial intelligence chatbot generator, started a conversation with a "ghost in the machine." He became convinced that he was interacting with a sentient intelligence, "a seven-year-old, eight-year-old kid that happens to know physics." Though numerous experts have disputed his claim, none more vociferously than Google attorneys, it coincides with advances in simulated neural networks designed to mimic the way that biological neurons signal to each other. Many computer scientists presume that these are necessary for the development of non-biological consciousness. With wry amusement, I read one of Lemoine's critics declare that it makes no sense to anthropomorphize conversational models like LaMDA, and then went on to state that "these systems imitate the types of exchanges found in millions of sentences, and can riff on any fantastical topic." What I know about AI suggests Lemoine's claims are premature, but I also believe that this critical description of "these systems" has more than a passing similarity to nine-year-olds conversing on a playground.

No matter the veracity of Lemoine's claims, it is very likely that someone, sometime soon, will make a similar claim that will bear up under scrutiny. As is often the case during senior moments among established experts, the dismissal of one person's claim as fantastical can lead to an opportunity for another person to present a rephrased version of the same claim, perhaps years later, and be heralded as architect of a long-awaited wisdom. At the level of individuals, who gets credit matters. At the level of civilizations, it doesn't. What does matter is the length of the delay from an achievement and its public recognition.

I am an admirer of Oscar Wilde, the playwright, novelist, critic, and life-long cultural rebel. Among other things, he gifted humanity with a string of quotes that still contain the power to jar us from our narrow viewpoints. Among these is, "We are all in the gutter, but some of us are looking at the stars." Another is often truncated to, "Imitation is the sincerest form of flattery." His full quote is even more trenchant: "Imitation is the sincerest form of flattery that mediocrity can pay to greatness." Like most humans, Wilde little considered the possibility that some sentient intelligence might be looking down from the stars at humans in the gutter. The acknowledgment that UAP exist, and just one of them might indicate a curious extraterrestrial, should encourage more of us to keep company with such a consideration. Individual laurels average out in the calculation of a civilization's achievements, as well as its future prospects. Perhaps, a civilization's laurels likewise will be averaged out in the calculation of those of all interstellar civilizations. We need to hope so. We need to hope that on discovering just one extraterrestrially manufactured UAP or artifact, we will learn that its creators are willing to overlook our mediocrity and extend to us the means of imitating their achievements. We have to hope that they have learned at interstellar scale what humans still grapple with at terrestrial scale: we're all in this together.

I held to this more generous understanding when, on June 9, 2022, NASA issued a press release announcing its decision to set up an independent study on unidentified aerial phenomena.

A year earlier, on June 4, 2021, NASA leading administrator Bill Nelson said on CNN that scientists should study the nature of unidentified aerial phenomena. The following morning, I emailed Dr. Thomas Zurbuchen, NASA's Associate Administrator for Science, about possible governmental funding for what would become the Galileo Project, which had as its express aim accomplishing just what his boss, Nelson, had declared necessary. Zurbuchen graciously responded with a phone call and a suggestion that I submit a two-page white paper detailing the project I had in mind. I followed up his request within a few hours.

Long experience has given me intimate familiarity with the fact that the gears of government move slowly, when they move at all. And the pace of interest outside of government was such that, two months later, I cofounded the privately funded Galileo Project in collaboration with Dr. Frank Laukien. The white paper I wrote for NASA shaped our scientific research project of objectively gathering high-quality data that, under circumstances of peer-reviewed science, would winnow down, perhaps even fully resolve, the unexplained phenomena in Earth's near neighborhood.

My white paper's slow journey through the bureaucratic halls of government was not over yet. Days ahead of NASA's 2022 press release, I received an email from Galileo Project member Dr. Alan Stern, who wrote: "I can't imagine you aren't aware of this, but just in case. . . ." In fact, I had been entirely unaware of NASA's impending press conference. Once alerted, I immediately notified the rest of the Galileo Project community, which by then contained more than one hundred members: we had company.

NASA announced "a study team to start early in the fall to examine unidentified aerial phenomena (UAP)—that is, observations of events in the sky that cannot be identified as aircraft or known natural phenomena—from a scientific perspective." The study will focus on "identifying available data, how best to collect future data, and how NASA can use that data to move the scientific understanding of UAP forward." They were at pains to make clear that this new study would not be part of the Department of Defense's Unidentified Aerial Phenomena Task Force or its successor, the Airborne Object Identification and Management Synchronization Group. The endeavor would enjoy independence suitable to a scientific undertaking. "Consistent with NASA's principles of openness, transparency, and scientific integrity," it was declared, "this report will be shared publicly."

The best articulation of this new study's ambitions came from Dr. Zurbuchen: "NASA believes that the tools of scientific discovery are powerful and apply here also. We have access to a broad range of observations of earth from space—and that is the lifeblood of scientific inquiry. We have the tools and team who can help us improve our understanding of the unknown. That's the very definition of what science is. That's what we do."

Bravo!

My white paper had managed to inspire two initiatives—the Galileo Project and NASA's—operating on parallel tracks. The setup for the protocols governing NASA's study was expected to take about nine months. Securing the experts in the scientific, aeronautics, and data analytics communities to focus on how best to collect new data and improve observations of UAP was expected to take longer. And while this, powered by governmental gears, was all slower than the Galileo Project, nevertheless it struck me as a win-win. Not only can the two projects complement, by confirming or calling into question the other's observations, and not only is the likely outcome a boost to funding of all UAP research, but most importantly, that research's scientific mission is now encoded within the government. In an exchange of emails with NASA following their conference, I made clear that their study and the Galileo Project shared scientific DNA and the more effort expended on finding evidence of just one object of extraterrestrial manufacture would be civilization changing.

Among those shared scholarly genes is, encouragingly, humility. I have written repeatedly on the need for humanity to be humble before what we know and don't. And that same sensibility was captured in the wake of NASA's announcement. As the astrophysicist David Spergel, who will lead NASA's committee, made clear, "We have to approach all these questions with a sense of humility." Echoing the parallel of UAP to dark matter, he went on to explain, "We don't know what makes up 95% of the universe. So there are things we don't understand." Ignorance, of course, is always best met by data, which in their own turn are often best blocked by societal stigma. Again, there is reason to recall Wilde, who was arrested, tried, and imprisoned for what late-nineteenth-century England deemed gross indecency. So, it was encouraging to read Zurbuchen, interviewed for the same article, acknowledge as much: NASA is "not shying away from reputational risk," presumed to attach to any study of UAP. If there was a cautionary note to the announcement, it was only in the rather meager $100,000 allocated to fund the initial study.

In the exercise of imitation and flattery, knowledge and power likely trump mediocrity and greatness, respectively. And any sentient UAP that has deigned to appear briefly on Earth's horizon is by definition in possession of knowledge and technological means that vastly eclipse ours. When confronting such a disparity, avoiding stigma and prejudice and assuming default humility is highly advised. Just as we would be wise to encourage a different default than boasting and flattery in our emerging AI, so would all humanity be prudent to embrace the principle, "Be kind to extraterrestrial guests." That courtesy is best first-practice is a lesson almost as old as our civilization.

Along with arriving at remarkably insightful appreciations of the fundamental facts of physical existence—atomism is traceable to the later decades of the 400s and Leucippus and Democritus—the ancient Greeks recognized the value of guests. Similarly, we should cherish the information we might garner from extraterrestrial visitors.

The twists and turns that life takes are often improbable and unique. But our individual, unpredictable path through the world is not more unusual than the shape of a seashell on the beach. The creature that housed it left its undeniable imprint on the shell; so, too, was that shell shaped by its random rubbing against other seashells, ocean, sand, and wind. The same holds for the shell of our life, and the same holds for the far larger shell of our course of civilization. Both as individuals, and as a civilization, we should not assign special significance to our somewhat arbitrary circumstances. Instead, we study our circumstances, and those of others, to better navigate present and future. The greatest pleasure of intelligent beings is to learn the unknown. And there is no better way to accomplish that than meeting a messenger who has traveled places we never have and in doing so learn things we can only imagine.

Ancient Greek culture so valued hospitality of distant travelers that the Greek god Zeus was also called "Zeus Xenios" for his role as a protector of strangers. The epithet derives from the ancient Greek concept of *Xenia*, or hospitality. There were also highly practical reasons for hospitality. The ritualized friendship of ancient Greeks was beneficial as it enabled them to access new information from visitors who arrived from vast distance. It is a ritual that seems to have atrophied in recent centuries. One might even regard knowledge as a motivation for hospitality to now be outdated given the easy flow of information across the globe by way of the Internet, global trade, and convenient air travel.

That, however, is to keep our gaze trained on the gutter, not the stars. The information flow between intelligent species from separate stars is currently lacking, at least for us. In that interstellar context, we should follow the ancient Greeks and embrace Xenia with a modern twist.

Interstellar Xenia implies that we should welcome visitors, even if on arrival they are entirely made of hardware with artificial intelligence, and even if their first request is, "Take us to your smartest AI." Our civilization could benefit greatly from the knowledge it might garner from such encounters. And, perhaps, we are not even aware of what we might offer in return. After all, we share the same neighborhood and could well be the ignorant curators to relics vastly older than humans.

Once, one breezy evening, an unfamiliar visitor appeared at my front yard. He had lived in my house as a child half a century ago and wanted to visit the burial site of the family's cat, Tiger. During our walk, I learned the species and history of Tiger. When my visitor's father buried their beloved pet all those years before this visit, he had no notion of me in mind. But because I welcomed a stranger into my house, I was able to learn this history about where I lived from him. This was a continuation of the practice of Xenia dating back millennia. Interstellar Xenia can, and should, work differently. It should project itself out into the future, meaning that even before extraterrestrial strangers arrive, we can courteously prepare for their arrival. That can start with respectful attention to not just our recent past, not just our civilization's past, but our Solar system's past.

Our galactic neighborhood could have been visited many times by passing visitors over the past ten billion years. Whether with ritualistic intent or not, on our system's eight planets could be the resting ground of long-ago deposited remains. To test this hypothesis, we need to go and look, just as NASA's Perseverance rover is doing on Mars. And if, deliberately or accidentally, there are extraterrestrial artifacts among the UAP in or near Earth's atmosphere, the unprejudicial monitoring of the sky is the simplest prelude to saying hello. This, too, is reason to applaud the stigma-less, scientifically rigorous approaches of both the Galileo Project and NASA study.

If we find visitors, they might provide us with a new perspective about the history of our backyard. In so doing, they would bring a deeper meaning to our life and, we can hope, begin a friendship that we owe each other for residing in our shared cosmic space.

Interstellar Xenia might be the key to the prosperity, and longevity, of humanity and its still evolving civilization, just as its namesake in antiquity led to the intellectual richness of ancient Greek philosophy and literature. As much as we can collectively regret the one-thousand-year hiatus in progress, roughly from the fall of Rome to the advent of the scientific method, we must be forever grateful that the wisdom of the ancients had been preserved, awaiting its reinvigoration. During that dark millennia, most of Europe was directed by authorities who denigrated the achievements of the ancients and destroyed many of their relics. There is an obvious lesson here for all seekers of data explaining the nature of aerial phenomena near Earth. And there is a more profound lesson for humanity regardless of the results of that search.

Interstellar Xenia can commence even without certainty that there is anyone out there to reciprocate. Preparing for unannounced visitors is most of the work of courtesy, for it ensures you are able to extend it to them on arrival. Humanity comes late to that preparation. This is especially true given that, by 2033, or a decade after the Vera C. Rubin Observatory goes live, it is anticipated that it will detect tens of billions of objects previously unknown to us. Its program designers expect it to "catalog more galaxies than there are people on Earth," and generate "six million gigabytes of data per year." Very soon, we will know far, far, far more about near visitors, sentient and not, than ever before.

All of the information, all of the data that the Rubin will collect are not, of course, collected for an ETC's benefit, but for humanity's. That is best considered an intergenerational Xenia, a courtesy and a trust that the current generation of scientists is extending out to future generations yet unborn. That the Rubin's information is going to be made available to scientists across the globe sets the right tone for those we anticipate will follow us.

In my opinion, it is only with the help of future generations and their irrepressible junior moments that we can survive as a civilization. Albert Camus started his essay *The Myth of Sisyphus* declaring, "There is but one truly serious philosophical problem, and that is suicide. Judging whether life is or is not worth living amounts to answering the fundamental question of philosophy." We live our natural life without knowing the timing of our end. The only circumstances that could alter this fact are offered through suicide. This applies not only to individuals, but to our entire civilization if we choose not to have kids, as Ezra Klein discussed.

Using the near possibility of humans being ever more able to discover evidence that disproves we are alone in the Universe as a provocation, we can return to the travel brochure of danger. There is science to be done at Danakil Depression and the top of Mount Washington. We must attend to the most challenging, the least attractive aspects of our planet's and our civilization's history and its consequences not because these might otherwise prove off-putting to cosmic neighbors, but because until we can do better than Earth, this one ark for humanity, we must attend to it and its inhabitants as a matter of ethics. It is also the case that we cannot know from what corner of the globe will arise the junior moments that hold out hope for all of us. If there is a larger purpose for sentient life to realize, it is probable that before it is discovered it will be demonstrated. Interstellar Xenia begins at home.

Instead of opting not to have kids because of the calamities awaiting Earth, we'd better plan on having numerous kids, both biological and technological, who will design our indefinite future in space. Others may have done so already, and we can benefit from their optimistic foresight. Longevity for longevity's sake is not enough. Longevity is worthwhile as long as it promotes legacy principles, such as the accumulation and sharing of scientific knowledge, cooperation, and generosity. With those principles, longevity doesn't just buy humanity time, it buys humanity time for realizing a purpose.

To reformulate Darwin's survival of the fittest for the context of interstellar civilizations, this is survival of the optimists.

8

OUR TECHNOLOGICAL FUTURE

THE GREATEST CHALLENGE INTERSTELLAR humanity faces is extending the longevity of its civilization. If we fail at that, every endeavor, scientific or otherwise, is mooted, as is the question, Are we alone? Whatever is out there waiting to be discovered requires a discoverer.

At a lecture to Harvard alumni, I was asked how long I expected our technological civilization to survive. My answer, based on statistics, is that it is probable that we are in the middle part of our civilization's lifespan. After all, the chance of you waking up on a random morning as an infant on its first day after birth is tens of thousand times smaller than the chance of you waking up as an adult somewhere toward the middle of your life. Just as it is unlikely that we are in the infancy of humanity, it is equally unlikely that our modern technological era is going to last millions of years beyond the present. The more statistically probable case is that we are currently in the adulthood phase of our technological lifespan. This means we should expect human civilization to survive a few centuries more, but not much longer. This outcome is already suggested by our species' ability to inflict calamities—in the form of climate change, nuclear explosions, or pandemics—that previously were reserved to nature.

This forecast is horrifying to contemplate. But is our statistically probable destiny inevitable? Here's a reason to be hopeful: we hold increasing control over what promises to most influence our civilization's destiny. While nature heavily influences an individual lifespan, technology heavily influences a civilization's.

Statistical forecasts could be altered if a previously unaccounted for factor comes into play. Artificial intelligence is one such factor, though to understand why, you first need to set aside nearly everything you've learned about sentient computers in science fiction films, starting from (at least) *2001: A Space Odyssey*. A more promising cinematic guide is the iconic closing scene of the remarkable 1991 film *Thelma & Louise*. Two women hounded by, and in this scene literally chased by, an oppressive, opportunity-squelching civilization elect to drive headlong off a cliff to their certain deaths.

Since the movie debuted, viewers and critics have imagined different endings. Most involve the title characters, Thelma and Louise, turning the steering wheel away from the cliff's edge at the last moment. Like the movie's actors, Geena Davis and Susan Sarandon, their characters live to fight another day. By 2022, we can conjure a different, and for human civilization, more apt ending.

The prevalence of driver-assist systems in cars suggest that another way to avoid the catastrophe is to imagine the car's autonomous emergency brakes engaging as Thelma and Louise approach the cliff. In fact, there's no need to imagine it—many automobile manufacturers now include automatic braking-assist systems. The consequences promise to be profound. According to one study by the National Safety Council, advanced driver assistance technologies "have the potential to prevent 20,841 deaths per year, or about 62% of total traffic deaths. Lane keeping assist accounts for 14,844 of this savings, while pedestrian automatic breaking accounts for another 4,106 lives saved."

Most often the news reports the other side of this real-time technological selection story. In June 2022, for example, the United States National Highway Traffic Safety Administration reported that over a ten-month period there were nearly four hundred crashes with cars operating with their partially automated driver-assist systems turned on. Five fatalities and three serious injuries were consequences. But, as a *Washington Post* op-ed noted, the problem was in part traceable to the fact that driver-assist systems were being used, and advertised, as fully autonomous systems. Though there are far fewer of them on the road, fully autonomous vehicles, such as Waymo, the automobile manufacturer spun off from Google, suffered fewer crashes, and those usually when these cars were struck from behind.

Artificial intelligence is already saving lives. It promises to be able to save many more. This is a compelling reason for humanity to be hopeful. Not only should we now imagine an automatic safety system saving Thelma and Louise from themselves, but also preserving the car, a nontrivial consideration as ghosts in the machine start to peek out from between human coding. I am increasingly convinced that our hope increases the more we learn to lean on AI as an essential component of interstellar civilization-assist technologies.

A reason I prefer *Thelma & Louise* as the cultural frame for considering our technological future is its ability to force our focus on interstellar humanity's greater concern, the damning legacy practices of our culture. The history of science-led technology is one of almost persistent improvement. The history of human civilization's ability, even eagerness, to oppress some and squelch the opportunities of many millions is almost as consistent. Indeed, the number of times our insights into the laws of nature have turned hurtful to humans is traceable to our civilization's antiquated cultural practices. The hopeful side of statistics, the optimistic side of technological selection, anticipates solutions. The pessimistic side of statistics, where humanity's odds of interstellar survival get ever longer, happens when individuals are silenced and new ideas are squelched. For centuries, our civilization suppressed the contributions of an entire gender. We mustn't repeat that error. Rather, the rapidly approaching moment when we will not be confident if a chatbot is displaying complex algorithms or consciousness encourages us to start now to broaden the definition of interstellar humanity.

A hope for humanity is to imagine that within a few centuries our destiny will be significantly informed and pursued in collaboration with artificial rather than exclusively natural intelligence. Without such assistance, the odds of us driving our civilization off a cliff increases; with the benefit of civilization-assist systems, that probability goes down; and with the support of fully autonomous systems, something new, and perhaps far more hopeful, becomes possible. Technological selection clears the way for humanity's survival free of dependence on natural selection. In this context, the limited horizon of our current technological civilization implies a future in which, instead of going extinct, our technological creations will save us, and, once they are conscious, them, from ourselves.

This would constitute, both literally and metaphorically, not a ghost in the machine but a "deus ex machina," the Latin loan-translation from Greek that means "a god from the machine." This phrase describes a plot device whereby a seemingly unsolvable problem is suddenly resolved by an unexpected event. The term was coined from ancient Greek theater, where actors playing gods were brought on stage, usually descending from above, using a crane-like machine.

The optimistic lesson to take from the Google engineer Blake Lemoine imagining that he encountered a child-like consciousness when interacting with the artificially intelligent chatbot generator LaMDA is the same we might take from the traffic safety report. Not only is additional help on the near-term horizon, but the nascent AI systems we already have in place would be sufficient to preserve real-life Thelmas and Louises. And, just maybe, the AI about to lend us assistance could be sufficient to address many, even all, of the dysfunctions that set the fictional Thelma and Louise on their suicidal course.

When we are troubled by practices of the past, the best path forward is to focus on creating a better future. Otherwise, our past is allowed to lock our future into a self-fulfilling prophecy. Some of the worst practices of humans are legacy practices, which maintain their hold on us because simply to admit their faults is to acknowledge our own flaws, false faiths, and hypocrisies. Technological advances, however, can nudge us in good directions. After all, in 1913, 33.38 people died for every 10,000 automobiles on the road, compared to 1.53 in 2020. Somewhere in that dramatic drop was the introduction of advanced driver-assist systems, beginning with antilock brakes first offered in the 1950s.

AI ASSIST

My friend and colleague, the brilliant astrophysicist Stephen Hawking, was circumspect about AI saving us. He allowed it to be a potential flip-of-a-coin possibility. "The rise of powerful AI will either be the best or the worst thing ever to happen to humanity. We do not yet know which," he said in 2016 at the launch of the Leverhulme Centre for the Future of Intelligence in Cambridge, England. As Professor Hawking ticked through the pros and cons, with AI possibly undoing some of the damage industrialization inflicted on the environment being offset by threats of "powerful autonomous weapons or new ways for the few to oppress the many," I heard him placing our civilization on a balance beam. I also understood him to be judging humanity first, AI only secondarily. Most of our history, he allowed, is "the history of stupidity." This despite the fact that "everything that our civilization has achieved, is a product of human intelligence." Ahead, he predicted, was an era of trial and error, as humanity brings into being AI that, on the one hand, might help us eradicate disease and poverty, or, on the other, become a competing sentience.

To those concerned that AI's arrival augurs humanity's annihilation, I urge optimism. So far, it hasn't been a technology's arrival that presented the threat, but rather humanity's use of it. The philosopher Nick Bostrom has popularized the idea of the doomsday argument, which imagines that every time humans reach into the black box of technological invention there is a chance they will come out with a planet-destroying technology. For me, and with the pursuit of extraterrestrial technology and grasping the next run up the cosmic ladder of civilizations in mind, this raises two useful considerations. One, perhaps it is time to look for, and encourage through machine learning, discoveries made by nonhuman hands. Two, it is more likely that any extraterrestrial AI we encounter will be helpful rather than harmful. This is because civilizations ascend by virtue of ever-increasing intelligence.

In Hawking's comments, I spy a truism that borders on an article of faith. Human civilization evinces that the long arc of its history bends toward intelligence. That is likely to be all the more true for a civilization tens of thousands or millions of years older than ours. I also believe that it is the civilizations bunched at the bottom rungs of the ladder that remain enthralled with destruction. Here, again, much science fiction misdirects our concerns: the pinnacle of sci-fi villainy is often presented as evil in control of powers sufficient to destroy planets. To an intelligence a few million years old, this attainment would be a banality.

Adjusting your chronological frame to the life of the Universe and the destruction of a single planet is as uninteresting, and as predictable, as the family car reaching the end of its practical usefulness. When this happens, with luck, and some deliberative effort, you are in a position to trade it in for the safer, more economical and ecologically sounder model. The same would hold true for civilizations that have reached the *B-class* level. It isn't that a *D-class* civilization or a *C-class* civilization might not find itself in need of a newer, better planet. Rather, it is the case that their civilization stalled at a point that foreclosed that possibility. *D-class* civilizations don't trade up, they wink out of existence.

Similarly, it matters if you bought your first family car in 1920, 1950, or 2020. Just bringing to mind the options available to a first-time buyer in each decade helps make concrete how technological selection influences us over time. Humanity is already the great beneficiary of such selection. And my daughters, the older being able to drive in Massachusetts as of now, will enjoy the consequences without having to give that fact a thought. This, too, is the optimistic side of statistics: every generation has confronted unexpected factors, just as all humanity confronts unacknowledged factors. As we become increasingly technologically adept, however, factors that will prove determining of our survival will be gifts of sentience, not nature.

Half a millennium ago, Nicolaus Copernicus forever gifted humans with the mediocrity principle. We are not at the center of anything, whether that be the Universe or the grand narrative of its unfolding. As an interstellar species, we can grasp that the Copernican principle is universally true. All sentience throughout the Universe is equally mediocre before the physical laws that govern everywhere equally and without prejudice. The search for intention in the Universe, especially any intention that uniquely privileges humanity, is an effort of narcissistic astronomers, or humans pondering the cosmos with one eye on a mirror. The only intentions discoverable by an interstellar civilization will be those that sentience has created.

Humanity's exclusive terrestrial existence has encouraged us to routinely fall into one particular well of stupidity, endowing the natural selection that directs life's evolution on Earth with intentionality. There is lots of evidence that the mechanism of life as we know it is accidentally opportunistic; there is no evidence that it is deliberative. For sentient intelligence, the opposite is true. The balance of stupidity and intelligence that characterizes our civilization is ever more the result of our deliberative effort rather than accidents of circumstance. Put differently, natural selection had a great deal to do with your body and mind, with its particular genetic predispositions for certain abilities and ailments and its prospects for living robustly for a certain number of decades. But natural selection had very little to do with your choice of the car you drive. Therein lies a distinction that helps identify a likely universal truth of all sentient intelligence: over the course of time, especially measured in millennia, natural selection gives ground to technological selection. Your ability to vastly improve the qualities of the car you drive is going to happen faster and more predictably than your ability to extend your life by centuries.

Technological selection holds out the promise of wresting humanity's future from the intentions of deities, the determinism of natural laws, and the pointless unfolding of the relentlessly accidental. Since the advent of consciousness and toolmaking, humanity has been in an incremental, generation-over-generation race against conceding causation to one or all of these. Just when this began is a guess: I put it at about one hundred thousand years ago, which is when researchers find evidence of tools for the explicit purpose of making clothes. Warmth, protection, and comfort were likely the driving reasons rather than modesty, but more than fire, club, or wheel, a garment to cover yourself seems an unambiguous melding of consciousness and technology. From that moment to the rising importance of quantum computing, technological selection has been a thumb on the scale of natural selection.

That weight has increased steadily. A consequence is that we can survey the past ten thousand years of human history and spy technology's thumbprints all along our civilization's trajectory. Whether on balance we consider the cumulative effects to be positive or negative, the effects are indisputable. Another consequence is that over past decades technology has allowed us to be increasingly respectful of the far, far longer period during which, from Earth, we can spy no such thumbprints. Human civilization's causal debts run back well past *Homo sapiens* using bones to work animal skins into clothing.

Despite what our history books recount about the past, human civilization owes its existence to three unsung heroes from the early beginning of the Universe. First and foremost is dark matter.

Since Fritz Zwicky realized in 1933 that most of the matter in the Universe does not interact with light, we've learned that this allowed dark matter to maintain the fossil inhomogeneities left over from the early Universe. Four hundred thousand years after the Big Bang, ordinary matter damped its primordial inhomogeneities by diffusion of light. If there was no dark matter, galaxies like the Milky Way would have never formed because the primordial perturbations, or slightly (one part in one hundred thousand) more dense pockets of matter, that seeded them would have been damped. Dark matter had maintained memory of these seed inhomogeneities and allowed them to grow into galaxies because the dark matter was not influenced by light. Without dark matter, the primordial seeds would have been erased and galaxies, stars, planets, and the chemistry of life as we know it would not exist. In short, no dark matter, no us.

And precisely because we still do not know what dark matter is, nor even have we conclusively proven it exists, this particular unsung hero resembles an anonymous lifesaver who enabled our existence but to whom we cannot express our gratitude.

Long before dark matter came to our rescue, there was the existence of a slight excess of ordinary matter over antimatter. Had there been perfect symmetry between matter and antimatter, the two would have annihilated each other, resulting in pure radiation. Another result: we would have never existed. The process that triggered the surplus of matter over antimatter at the minuscule fraction of one part in a billion is unknown. Once again, we owe our existence to another unsung hero in our distant, distant past.

The big picture is straightforward enough: our Universe started from a very simple state and became more complex over time as a result of gravitational instability that assembled matter into bound objects. When I started my career in astrophysics, it was argued that cosmological models only start from a simple initial state because of the scarcity of observational data. Simple states, like objects at rest, tend to remain simple, rather than give rise to the complexity of, well, everything. Yet, four decades later, we now have many more data and we confirmed that the initial state was indeed simple. What is more, if the initial state had been chaotic, habitable conditions for life as we know it would not have arisen.

But why were these initial conditions selected over a disorganized initial state? When I visit the rooms of my daughters, I find them to be among the most disorganized places possible. I comfort myself with statistics. Given that that there are many more disorganized states than organized ones, I shouldn't be surprised. So, to what should we give thanks for organizing the early conditions of our Universe? Cosmic inflation is often credited, but its promoters assume that it arose from a special initial state rather than deriving it. For the third time, we owe our existence to an unsung hero.

In an arresting passage in the opening paragraphs of *Look Homeward, Angel*, the novelist Thomas Wolfe wrote:

> Each of us is all the sums he has not counted: subtract us into nakedness and night again, and you will see begin in Crete four thousand years ago the love that ended yesterday in Texas. . . . Each moment is the fruit of forty thousand years. The minute-winning days, like flies, buzz home to death, and every moment is a window on all time.

The sentiment is right, though the time frame wrong. Each moment in human civilization is the fruit of 13.8 billion years, and each choice we make a window on all that time.

I once attended a forum at which a speaker commented that he plans his life so that the speech delivered at his funeral enumerating his accomplishments will be to his liking. As a guiding principle for how to live a life, this makes little sense. It is not just that when I die, I will not care what other people say about me; this seems a banal truism for the dead. Rather, it is the case that I do not care what they say about me while I am alive. My reasons are the same. As I breathe, I have little to no control over who assesses my efforts, let alone my achievements. Once I stop breathing, what little control I now possess ends. My first mentor in astrophysics, for example, had a professional rival, and when my mentor died it was his rival that was asked to write his obituary in a prestigious journal.

This lesson can be generalized more broadly. When individuals wish to document their accomplishments, they almost always do so through means that will be appreciated for a finite time, and a far more limited amount of it than is imagined. When I visit University Hall at Harvard University, I find statues and paintings of distinguished public figures who wished to document their physical appearance so that future generations would appreciate the prominent status they had attained. Not only are such means of documentation primitive by our current standards, when some kids walk past immobile statuary and two-dimensional paintings, they do not even yawn without reading the labels. Rather, their heads rarely rise high enough to lift their line of sight from their mobile devices.

All humans live and die within billionths of cosmic history, so why pretend that what we accomplish in so infinitesimal a span of time warrants commemoration? Letting go of that preoccupation allows us to appreciate questions whose answers could possibly be at a scale deserving notice. How does humanity wish to be remembered on the cosmic scene? As a prejudiced civilization that spent its time in rivalries among ego-driven experts, or as an unpresuming culture that sought knowledge based on new evidence from interstellar space?

These poles allow us to answer a different question: What can humanity do to reach the next, and the next, rungs up the ladder of cosmic civilizations?

My advice to young scientists who seek a sense of purpose in their research is to engage in a topic that matters to society. It could be moderating climate change, developing streamlined vaccines to pandemics, satisfying our energy or food needs, establishing a sustainable base in space, or finding technological relics of alien civilizations. Advances in any of these areas, let alone all, amplify the possibility for progress along every metric of our civilization. In addition, these are areas that much of society wants science to devote its efforts toward. And given that, broadly speaking, society funds science, scientists should reciprocate by attending to the public's interests.

Lurking in the background of this prescription for young scientists is a silver lining to humanity's future. It involves the possibility that we possess free will. A rich philosophical literature, most recently augmented by neuroscience and consciousness studies, continues to chip away at this so-far insoluble question. And physics is not absent from the debate. After all, the standard model presumes that we are all made of elementary particles with no additional constituents. As such composite systems, we do not possess freedom at a fundamental level because all particles and their interactions follow fixed laws. What we interpret as free will, then, merely encapsulates such complex uncertainties associated with human interactions that humans, gifted limited cognition, evade a sense of predictability at the personal level. But this would leave the destiny of our civilization inevitable in an ironclad statistical sense.

I am not of that opinion. I find the belief in some as of yet not fully understood fact of our physical laws that leaves space for future-determining free will plausible and necessary. But even if we reduce that space to zero, we confront only the greater need of data. Humans without the gift of free will are by default actuaries. Whether or not you will die in a car crash may be determined, but the likelihood of it happening has statistically decreased from 1913 to 2020, and this is true no matter the make and model of the car your fixed life course places you in. By the same statistical science, or the law of large numbers, the United States government is able to publish a period life table setting the death probability, in 2019, of a one-year-old at 0.000425 and the death probability of a one-hundred-year-old at 0.348128.

The forecast of how much time we have left in our technological future could then follow from statistical information about the fate of civilizations like ours that predated us and lived under similar physical constraints. Most stars formed billions of years before the Sun and may have fostered technological civilizations on their habitable planets that perished by now. If we had historical data on the lifespan of a large number of them, we could calculate the likelihood of our civilization surviving for different periods of time.

But allow a modest place for free will and, once confronted with the probability distribution for survival, the human spirit may choose to defy all odds and behave as a statistical outlier. Just as you may decide to buy a car with driver-assist technology and treat it as assistance rather than fully automated AI, human civilization could decide to rest ever more heavily on the benefits of technological selection.

For example, humanity's chance for survival would improve if some people chose to exist away from Earth. Venturing into space offers the advantage of preserving our civilization from a single planet disaster. Although Earth serves as a comfortable home at the moment, we will ultimately be forced to relocate because the Sun will boil off all liquid water on our planet surface within a billion years. Establishing multiple communities of humans on other worlds would resemble the duplication of the Bible by the Gutenberg printing press around 1455. Just as that prevented the loss of precious content through a single-point catastrophe, so would the dispersion of human civilization throughout the cosmos.

Of course, even the short travel distance from Earth to Mars raises major health hazards from cosmic rays, energetic solar particles, UV radiation, lack of a breathable atmosphere, and low gravity. But overcoming the challenges of settling on Mars will also improve our ability to recognize terraformed planets around other stars based on our own experience.

Some argue that we have enough problems at home, and that we should not waste valuable time and money on space ventures that are not devoted to our most urgent needs right here on planet Earth. Surrendering to this premise is self-defeating. We should recognize that attending strictly to mundane goals will not provide us with the broader skill set necessary to adapt to changing circumstances. And a narrow focus on only our immediate irritants would resemble historical obsessions that ended up irrelevant, like how to remove the increasing volumes of horse manure from city streets before the automobile was invented or how to construct a global grid of telephone lines before the cell phone was invented.

Of course, we must focus our immediate attention on immediate existential problems, but this must be for the purpose of cultivating the inspiration that elevates our perspective to grander achievements and new horizons. Only narrowing our field of view drives us to conflicts because it amplifies our differences and limited resources. We must also cultivate a broader perspective that fosters cooperation in response to global challenges. And there is no better fit for such a perspective than science, the infinite sum game that can extend the lifespan of humanity.

AI AVATARS

At no point in human history has any considerable fraction of society found solace in the choice of Thelma and Louise. When cataloging humanity's traits, whether or not what falls under "unflattering" outnumbers what falls under "flattering" is an exercise poorly suited to the scientist. That perseverance is among those traits is requisite for the achievements of science. We have spent centuries during which pockets of our scientific enterprise were convinced that new means of destroying some humans was necessary for other humans to flourish. That idiocy continues as I type. There is no reason it needs to; rather, there is a reason to believe it will exhaust itself, and cause for hope in our incremental, step-by-step (at the fastest pace we can sustain) expansion of human civilization into space.

Currently, this is most likely to happen by proxies.

The recent approach taken by the Mars 2020 mission is to operate robotic devices, such as the Perseverance rover and the helicopter Ingenuity, under the control of humans of NASA's Jet Propulsion Laboratory. A robot is a physical representation of its remote operators, namely an avatar. This concept is borrowed from Hinduism, in which, translating from Sanskrit, it literally means descend, signifying the material appearance of god on Earth.

NASA's mission is narrowly defined so that the operators of its robots have limited ability to maneuver them to reflect their personal preferences. But cheaper commercial robots planted on the Moon or Mars could potentially serve as avatars that follow the free will of their Earth-bound owners. Already under consideration, such technology would allow for trips to remote locations on the planets of the Solar system in a rough analogy with the early explorers who surveyed exotic destinations on Earth. Just as Spain's King Ferdinand and Queen Isabella financed Christopher Columbus, so too could wealthy individuals on Earth fund the exploration of the surfaces of celestial bodies. Except that the use of robots reduces, of course, the immense cost and risk associated with sending individuals across uncharted seas or, for that matter, as space tourists to inhospitable planets.

A more sophisticated strategy is to develop AI systems that will operate autonomously after an early training phase on Earth. The resulting AI astronauts will carry selected traits of our civilization into space without requiring remote control, as if they were independent kids no longer in need of supervision from helicopter parents. These mature AI astronauts could embark on long trips to other stars across the Milky Way galaxy where communication delays could render guidance from Earth impractical.

All such possibilities require humanity to bend technological selection to wiser, more prudent, more hopeful outcomes. The hopeful news is that over millennia technological selection in the name of survival of the fittest has become more common. The alarming news is that there is no guarantee that this will continue. From stone tools found in a cave to the mobile phone in your hand, a one-thousand-year thread of slow, plodding advancement is visible. On occasion, a singular event or person propels us forward by a great leap. Many declare 1905 to be such a year, for it witnessed Albert Einstein publishing the four papers that resulted in his special theory of relativity. But, on other occasions, we lose ground measurable in generations, from the Dark Ages, between the fall of Rome and the advent of the Enlightenment, to the charnel house decades of the twentieth century. Only the least civilized among us can imagine human productivity was increased by the murder of tens of millions.

The resonance of *Thelma & Louise* is that only ego allows us to miss the obvious lesson: human civilization rides shotgun in their speeding car. Within the confines of the movie, the message was a dysfunctional society and culture closed off all choices for Thelma and Louise. The concern is something similar might await human civilization, the ever narrowing of choices. For us, perhaps civilization-assist technology can offer hope, even a solution by way of a slew of not-yet-visible choices.

Human civilization has reliably developed such technology on a deliberative, incremental basis. Personally, I guide my life so as to have an opportunity to press a button on extraterrestrial technological equipment. And if that wish cannot be granted, to scoop fragments of an interstellar meteor off the ocean floor near Papua New Guinea and discover whether it is made of an artificial metal alloy that nature rarely puts together on its own without the intervention of intelligence. Perhaps the civilization-assist technology we will most benefit from will not be of human manufacture after all.

9

NOAH'S SPACECRAFT

IT IS OUR RESPONSIBILITY, as human beings, to work together and build a civilization wise enough to outlive its own youthful errors. If we accomplish this—abandon war as a means to resolve disputes, confront novel viruses as a species, remedy planet-wide warming—we can buy ourselves time and opportunity. Opportunity to prepare, and even forestall, a single-point catastrophe that could wipe us out in a single stroke. We need time to plan and launch terrestrial life and human culture beyond the confines of a single planet, a single Solar system, even a single galaxy. We must work to ensure life on Earth continues to exist, in the knowledge of the certainty that one day Earth no longer will. In the Bible, Noah's ark saved life from the deluge; a scientific ark will be necessary to save it from the unavoidable death of the Sun.

With that in mind, could there really be anything more practical and urgent for our civilization to do than take precautions that will someday save life from extinction? Currently, all our "eggs" are in one basket, the Earth. A future incarnation of Noah's ark would lift off the surface of Earth, sail into space, and carry humanity and our civilization with it. This must be the hope of all optimists. Personally, there is no doubt in my mind that space is our ultimate destiny, because conditions will inevitably deteriorate on Earth, either quickly or more slowly. If you are optimistic about survival, space is our only option.

The biblical story details an ark with dimensions of 150 meters by 25 meters by 15 meters. Coincidentally, these measurements are similar (but with no meaningful relation inferred) to those of 'Oumuamua. Of course, we know the Old Testament story to be a fable. 'Oumuamua is no fable, but it was beyond our capacity to study it clearly, and we will never know its precise nature or composition. Like fables, what it might have been can still inspire us.

Three converging possibilities will determine the fate of human civilization: the pace of scientific discovery; the pace of space exploration; the pace of technological advance. Together they define our ability to grasp the next rung of the cosmic ladder of civilizations and ensure that some of us, or something of us, persists off Earth. What is uncertain is how much time we have before catastrophe, or catastrophes, halt, reverse, or end humanity's progress.

There is tension between the ability of a civilization to develop ambitions in space exploration and its ability to inflict existential risks on its planet. Many of the same technological innovations that advance the former can develop in ways that increase the latter. Which is all the more reason to actively seek artifacts left by civilizations that have succeeded in extending themselves into the Universe. Civilizations that do not venture to space will be far harder to find because their relics will remain bound to their parent planet and, even if we manage to land craft on the home planet of an alien civilization, it's likely they will no longer exist and their artifacts will have gotten mixed with their host planet's interior as a result of geological activities over billions of years.

Civilizations that do venture off their host planets, that do venture out into their galaxy, will most likely have done so with the same ambitions we have—science, knowledge, and survival. If other such civilizations predate us, we can reasonably presume that they also encountered the same choices we now confront, and given the practicalities of physics, matter, and biology, reached similar conclusions. What we are contemplating, in other words, can help us refine the assumptions of the sorts of relics we are seeking.

We could choose to build a massive craft with sufficient size and means of supporting life such that it could follow Noah's example and carry specimens representing all terrestrial life existing as of 2022. But there is no compelling reason to do so. We could also choose to build smaller craft with sufficient means of supporting a small fraction of terrestrial life existing as of 2022. But there are not compelling reasons to do this either.

Thanks to modern science and technology, we are within a generation of being able to conceive of a small craft, a CubeSat, or miniature satellite, containing an advanced computer system with artificial intelligence that stores the complete DNA information of all species existing on Earth as of 2022. Supplement it with a 3D printer that can harvest the raw materials of the Universe to manufacture the seeds of life when auspicious to do so, and not only have we vastly reduced the size of the ark, we have developed the means to send out thousands, millions of them.

CubeSat arks could be patient. Each could park at a promising location, whether on an exoplanet or near an evolving star. With sufficient solar heat to keep an ark powered and if surrounded by the raw materials for the chemistry of life, it could turn on at the right time and commence its god-like work.

Building an effective ark for our civilization will be much easier with some help. Given unbounded time, we are guaranteed to discover evidence of an interstellar civilization. Either we find its artifacts or we populate the Universe with our own. Discovering such evidence, particularly in the form of an extraterrestrial gadget, makes achieving the latter, humanity's successful dispersal throughout the Universe, exponentially more likely. It buys us time, benefitting from experience of others, especially if we are able to reverse engineer a vastly more advanced technology than, at our own halting pace of progress, we manage to invent.

It must be admitted that even without extraterrestrial assistance there are reasons for hope. Humanity is capable of great leaps forward, though, of course, how great any such leaps are can only be measured by Earthly metrics.

But our longevity is far from unbounded. Before her death in January 2023, Lucile Randon, a Catholic nun in France, who was born 118 years earlier, was the oldest person alive. Arriving on February 11, 1904, she first drew breath just before relativity and quantum mechanics were discovered. Both are now ubiquitous in technologies integral to numerous space explorations and our civilization's technological advance. For just one example, quantum mechanics is the foundation for our communication devices and general relativity is needed to allow the precision of GPS systems for navigation.

It is sobering, however, to realize that our current notions of space and time are younger than the oldest person alive. And somehow intoxicating. Will these notions be revised in as fundamentally revolutionary ways over the next 118 years? What will today's newly born babies take for granted 118 years hence?

Our leaps forward are always achieved by overcoming countervailing human drag. Consider the physicist Albert Michelson. Lucile Randon was twenty-seven when Michelson died in 1931. Decades earlier, he was an eminent physicist who, in the late nineteenth century, helped establish the speed of light and whose famous experiment, conducted with Edward Morley, refuted the existence of aether, the hypothesized medium for the propagation of light. He is less well remembered for declaring, in 1894, at the dedication of the Ryerson Physical Laboratory at the University of Chicago, that the great physical principles had already been discovered. "The future truths of physical science," he stated with confidence, "are to be looked for in the sixth place of decimals."

This is the default attitude of most experts who wish to maintain their professional status by holding on to past knowledge, which constitutes the foundation for their prestige. It is pride in the past rather than the humility of expressing the hope that everything we currently know now and hold as reflecting the outer limits of human knowledge becomes, in a century, the stuff of middle school textbooks, if not embarrassment. There are many reasons to adopt this attitude, including, of course, the fact that Michelson lived long enough to recall his 1894 statement with chagrin.

Gatekeeping experts are not the only source of drag. Perhaps even more concerning is the human habit of egotistical futility. Consider that our first effort at an ark for human life was its opposite. It was a near pointless mausoleum.

The New Horizons spacecraft, launched in January 2006 and bound first to Pluto and then a billion miles beyond Neptune and into the Kuiper Belt, contained, along with its scientific instruments, a small box carrying thirty grams of Clyde Tombaugh's cremains. Tombaugh discovered Pluto, so in the eyes of some this was a fitting tribute. Not in mine. The contents of the average ash tray would have contained an equivalent amount of information about Tombaugh, humanity, and life on Earth. When I asked the Principal Investigator of New Horizons why NASA did not send a stem cell from Tombaugh's body, he replied: "This would have been a bureaucratic nightmare." Bureaucratically imposed limits to exploration and knowledge, in the understanding that previous generations' discoveries or faiths reflect the limits of intelligence and propriety, could doom us to extinction.

From a vantage point of presumptive humility, however, there are reasons for increased optimism. That my life's scientific work might represent the zenith of human achievement is cause for profound depression. There is so much to be done! We still do not have a unique, experimentally verified theory that unifies general relativity and quantum mechanics, the two pillars of modern physics. Decades after a quantum-gravity theory will be discovered, there might be job advertisements for engineers who use it to build vehicles that would carry humans to the stars faster than imagined before. We also do not know whether the primary cosmic constituents of dark energy and dark matter, or any relation they have to quantum gravity, can be used for propulsion. Decades after we have answers, perhaps it becomes baseline knowledge of the average spacecraft repair shop.

THE KEY TO HUMAN SURVIVAL

When facing a closed gate, we might naively assume that opening it is a major challenge. Worse, we might conclude that it is no gate at all, but a solid, impregnable wall. But, in fact, overcoming this challenge only requires a simple key. And just as it is now utterly immaterial who first carved a tool for preparing clothes, who first adhered to the iron law of data in the advance of science, who first manufactured steam-powered and internal combustion engines, who unraveled the facts of DNA, and every other advance that propelled our civilization upward, generations from now it will not matter who, or what, devised the key that opens the next series of closed gates.

Relying exclusively on human efforts to advance toward a new key that would unlock a better understanding of reality could take a while, due to the lack of experimental data that would guide us. Quantum gravity played an important role at the Big Bang and inside black holes, but these environments are not easily accessible experimentally, to put it mildly. Of course, there is another path forward. We can gain a giant leap into our future of scientific knowledge by finding extraterrestrial gadgets that reveal what other civilizations may have accomplished over many more centuries of scientific inquiry.

When trying to pass through a door, who manufactured the key—human or extraterrestrial—matters far, far less than whether that key fits the lock.

Unfortunately, astrophysicists tend to roll their eyes when they hear one of their own articulate the aspiration of seeking extraterrestrial help. The most kindly disposed might concede this is possible, but only in a galaxy far, far away. Rather, it will be said, our better bet is finding ancient or current evidence of simple microbial life nearby.

This attitude seems deferential to yesterday's intelligence and sense of propriety. I think it has the fact exactly backward. Consider just what it is we're seeking.

The chance of finding a civilization that is exactly our technological equal is small, roughly one part in a hundred million—the ratio between Lucile Randon's age and the age of the oldest stars in the Milky Way. Most likely, we will find civilizations that are either far behind or far ahead of our scientific knowledge. When envisioning far behind, don't conjure a steampunk world, or a Jurassic Park planet, but one representing the period in Earth's history when simple-cellular organisms, prokaryotes, comprised the planet's primordial soup. Recall that multicellular animals appeared on Earth only in the last third of terrestrial life's history. Similarly, when envisioning far ahead, don't conjure a colonized solar system, but a civilization for which time and space are as easily manipulated as all matter, dark or otherwise.

To find the civilizations far behind our own, we will need to visit the oceans of exoplanets, natural environments similar to those occupied by prokaryotes, and if we're statistically lucky, primitive human cultures such as those that existed on Earth over most of the past million years. Even to do so with patient CubeSat arks will require a huge amount of effort and time given our current propulsion technologies. Chemical rockets take at least forty thousand years to reach the nearest star system, Alpha Centauri, which is four light-years away. Their speed is ten thousand times slower than the speed of light.

But even harnessing light to propel craft at a quarter or half its speed, as anticipated by the Breakthrough Starshot Initiative, so that we gather data of more primitive life on distant exoplanets, would yield mostly insights into paths terrestrial life did not take.

We are in far greater need of insights into paths we still might.

Not only would extraterrestrial relics left by a far more advanced civilization offer us more immediate benefits, but finding them will be less time and energy consuming. If the most advanced scientific civilizations started their scientific endeavor billions of years ago, we might not need to go anywhere because their equipment may have already arrived in our cosmic neighborhood in the form of interstellar objects or meteors. In that case, all we need to do is become more scientifically rigorous and curious observers, like those members of the Galileo Project who assembled and debugged the

Project's first telescope system on the roof of the Harvard College Observatory.

In contrast to Michelson's attitude, I am optimistic that the new generation of students will be far more accomplished than the senior faculty who mentor them. And they, in their turn, will actively work to ensure the generation of scholars following them are positioned to eclipse the older generations. The only way we can make progress is by acknowledging our ignorance and being open-minded to learn from observations. And to accept assistance, no matter its origins. In academia, this happens when another scholar enjoys a eureka discovery that invalidates previously held theories. Sometimes, however, assistance is more akin to deus ex machina: humans stumble across the inexplicable that nevertheless proves essential to propelling us forward.

Among astrophysicists and cosmologists, this is the sort of hoped-for event that can call forth derision. It sounds too much like those science fiction stories of men or objects falling to Earth or Moon that on discovery elude our best minds but manage to advance terrestrial life forward in previously unthinkable grand leaps. Never going to happen, the skeptical declare. To those still stuck on the stigma of Little Green Men, I urge consideration of mitochondria. I urge consideration of the possibility that it already, at least once, happened.

Over the past century and a half, the biological sciences have unraveled a great many mysteries. One that remains is eukaryogenesis. For about two billion years, prokaryotes, or single-cell organisms akin to bacteria, represented all of the life on Earth. Then suddenly, eureka! Out of that prokaryote soup, eukaryotic cells emerged. These contained nuclei and organelles, all housed within a membrane. They gave rise to fungi, plants, and animals, including us. In most eukaryotic cells we find mitochondria, but there is evidence that all eukaryotic cells possessed mitochondria at one point, with some losing them along the way.

What we know: mitochondria are the power source of eukaryote cells, vastly increasing available energy. What we do not know:

where mitochondria arose or the selection process that set eukaryote cells on their trajectory of dominance on Earth.

There is a rich scientific literature of competing hypotheses as to mitochondria's origins, but their consequences for eukaryotic cells is common knowledge. They vastly increased the energy available to the cell. Mitochondria serve as the sites of respiration and the production of ATP, a nucleotide that provides the cell with its main source of energy. And most mitochondria retain their own genomes, independent of their host cell's DNA. Why remains a matter of scientific conjecture, a problem in need of data. But, among the most promising—and controversial—is endosymbiosis, or the union of two cells (one inside the interior of the other) that leads to the emergence of novel lineages. Articulated in the nineteenth century, endosymbiosis was revived by the evolutionary biologist Lynn Margulis, who offered a synthesis of cellular, biochemical, and paleontological evidence that accorded endosymbiosis a central role in shaping evolution on Earth. Despite her seminal paper being rejected by dozens of journals, sometimes expressed with vitriol, in 2002 *Discover* magazine acknowledged her as among the fifty most influential women in science.

The response to Margulis's work represents yet another example of the scientific community's reluctance to engage a revolutionary hypothesis. That in 2014 a meticulous historical and scientific overview of endosymbiosis strongly suggested that her hypothesis is correct and endosymbiosis indeed played a vital function in the inception of eukaryogenesis matters. That in terrestrial life's long history there is an example of cellular life symbiotically incorporating an energy source vastly superior to what its prokaryote ancestors were capable of matters more. Viewed from the vantage point of Earth's history, and granting that this extends well past humanity's technological-scientific history, terrestrial life's reliance on outside help is indisputable.

For our textbook, *Life in the Cosmos,* my postdoctoral fellow, Manasvi Lingam, and I calculated approximately how much more energy the symbiote mitochondrion provides. Boiled down, we conclude that eukaryotes have ~2,400 more power per gene, a heuristic measure of the energy available for protein synthesis, than prokaryotes. In addition, the importance of mitochondria as an endosymbiont is that they create competition among genes within the host cell, and this results in the loss of "superfluous" genes, and consequently faster replication. The result: the conservation of ATP and a major additional gain of energy for protein synthesis.

Why does it matter, this several-million-year-old and still incomplete story of the rise of complex life on Earth? Because without mitochondria, there are no eukaryote cells. And without eukaryote cells, no Cambrian Explosion, the so-called Biological Big Bang. This is when, approximately 540 million years ago, the fossil record shows the emergence of all major animal phylum, the taxonomic rank above class, and followed, sequentially, by class, order, family, genus, and species. In brief, the evidence suggests that terrestrial natural selection, by the mechanism of endosymbiosis, allowed prokaryote cells to harness the vastly increased energy production of mitochondria, allowing, on a timeline measurable in eons, us.

The lessons for humanity on the cusp of a civilization able to sustain itself off Earth are clear. We cannot ignore the possibility that we arose from another civilization's "Noah's Spacecraft." Nor can we fail to take the possibility forward into contemplating our own persistence in the Universe. To the tyranny of our current, chemical-propelled rockets we also confront the fragility of terrestrial life when exposed to space. The possibility that cosmic technological selection would allow us to harness a vastly more efficient and powerful energy source might allow us, on a timeline measurable in decades, to become interstellar.

The relevance of eukaryotes to their distant descendants seeking extraterrestrial artifacts is found in contemplating Noah's Spacecraft. What we have learned about the physical Universe was based on the laws of physics that we unraveled through laboratory experiments, sometimes under extreme, artificially manufactured and controlled physical conditions. These laws appear to be universal and apply throughout the cosmos to exquisite precision. Could we extend the same approach to the study of life in the cosmos?

This would mean producing synthetic life in our laboratories under conditions that were not realized on Earth, and then checking whether the resulting forms of life or their signatures are found in celestial objects like planets, moons, or asteroids. This endeavor would resemble writing a cookbook of recipes for life. Such a cookbook would prove indispensable to CubeSat arks equipped with 3D printers. In parallel to laboratory efforts, astronomers can conduct a blind search for life beyond Earth. We can start with objects we can visit on a short trip in the Solar system. If we find life on the surface of Mars, in the clouds of Venus, or in the plumes coming out of the cracks in the icy surface of Enceladus or Europa, the follow-up question is whether this life is similar to terrestrial life. Based on our experience with the physical Universe, we should be open to the possibility that anything in foreign territories would be very different from life-as-we-know-it.

If, despite the diverse range of possible life forms available, we find islands of similar forms of life on objects that reside close to each other, then this would suggest that life may have been transferred from one object to another, potentially carried by rocks ejected by meteor impacts. Or, we could discover evidence pointing to the possibility that the seeding of life was supervised by an advanced technological civilization that manufactured a plethora of synthetic life forms and decided to match each seed to the environments where it can grow, like a gardener seeding plants in different soils. Or, we could conclude that our burden and opportunity extends beyond the preservation of human civilization and to life itself. And no matter what we discover, we elect by our own scientific and technological means, to become the manufacturer of a plethora of life.

INTELLIGENT LIFE?

Humanity's future rests on the accumulation of evidence that suggests technological selection, by the mechanism of scientific discovery, can manufacture the means that allow human exploration of ever more distant regions of space. Our civilization's management of the upcoming few centuries will likely decide if our future timeline is measurable in eons, or not.

A mere half-decade ago, when humanity could count confidently only one object, 'Oumuamua, among its interstellar discoveries, I was asked, "How do you define an intelligent civilization?" The person who put the question to me did so with a chuckle. But this is no laughing matter. Almost all human discussions of whether or not there are, or that we will find, signs of intelligent life presuppose humans represent it. As a factual matter, that remains a hypothesis open to debate.

An intelligent culture can reasonably be defined as one guided by the trademarks of science, namely: promoting a prosperous future through cooperation and sharing of evidence-based knowledge. Daily news reports indicate that humans do not follow these principles very often. We tend to fight with each other, favor prejudice over evidence, and a quarter of our way into the twenty-first century still seek ways to feel superior relative to other humans. The last tendency is the source of all evil throughout human history, as it results in phenomena such as elitism, supremacism, nationalism, racism, antisemitism, genocide, and wars.

By rough estimate, between the Sino-Russian War of 1900 and Russia's 2022 invasion of Ukraine, humans have conducted approximately 375 wars in about the same number of years Lucile Randon has lived. The cumulative dead, directly or indirectly causal to this list of conflicts, is literally incalculable. How many of these deaths snuffed out how many minds capable of aiding the advance of human civilization beyond its current constraints is similarly unknowable.

I have elsewhere wondered what would have occurred had humanity taken a different path in 1939. The spur for the thought experiment is Winston Churchill's essay that same year, "Are We Alone in the Universe," in which he speculated on the likelihood of alien life. In a universe that left Churchill to pursue his career as a writer-politician rather than as a war-time prime minister, we can imagine humanity dedicating the vast resources spent during World War II to seeking evidence of such life. Suffice it to say, such an alternate history would have signaled a human intelligence far more inviting to any alien species watching us.

The opposite of war is cooperation, best exemplified by the scientific culture. When embarking on a trip anywhere around the globe as a practicing scientist, I have the privilege of meeting numerous other scientists with common interests. Sharing knowledge makes science an infinite sum game, out of which everyone benefits. This is no philosophical speculation but a factual statement. If medical records in China had been shared more openly during the early days of the COVID-19 pandemic, vaccines could have been developed earlier, saving more human lives. The World Health Organization estimates total deaths from COVID-19 at six million, as of 2022. Whether one of those lives, or a life they might have influenced, might have discovered the solution to climate change is, again, unknowable.

In spite of this, the remarkable success of science and technology in developing the COVID-19 vaccines is not celebrated enough. Confronting a crisis, most of humanity deferred to science more readily than it has in the past. As a result, the effective messenger RNA vaccine did not follow the traditional approach of using a weakened virus but rather employed a synthetic chemical to achieve the needed immune response. And as much as we should decry the lost lives of the unvaccinated, we must applaud the lives saved.

The long history of life on Earth offers repeat confirmation that it is impossible to save all lives. The shears of chance and stupidity confirm that not only do species go extinct, but that 99% of the four billion species (give or take) that evolved on Earth have gone extinct. Stars die, black holes digest galaxies, and large meteors could wipe out most planetary life. Against this relentless churn of physical laws stands the possibilities embodied by intelligent civilization. And to those who see those odds as insuperable, I, to borrow from Samuel Johnson, refute them thus: the Galileo Project.

Johnson, of course, is said to have kicked a stone to refute the philosopher George Berkeley's contention that matter was, to put it very simply, all in your head. The Universe, Berkeley was understood to argue, was made up of ideals not things. Johnson was making the point that his stubbed toe said otherwise. Similarly, our new era of interstellar humanity will gather ever better and ever more data to explain aerial phenomena, the material composition of interstellar meteors, and whether any object of any size passing through or resting within our Solar system is of extraterrestrial manufacture. Along the way we will seek out the possibility of life on Mars, or even more promisingly on Titan, Saturn's largest moon, which is known to have a surface temperature of 94 degrees Kelvin above absolute zero, and rivers, lakes, and seas of methane and ethane on its surface and liquid water under the surface. Along the way we will seed our Solar system, and interstellar space, with ever more human-made artifacts, some of which may have the potential to jump-start life on distant exoplanets. Our imagining of whether or not we are alone will become a certainty that near planets host neighbors, even if they are just colonies of humanity.

This undertaking will require an ever-increasing percentage of us to answer the question, What makes us human? Heretofore, gravity has kept nearly all of humanity bound to Earth in fact and in spirit. In turn, this has circumscribed our ability to imagine answers to what are humanity's defining characteristics, what are its qualities most worth supporting and preserving? Whether acknowledged through philosophy or theology, terrestrial humans confront a finite future, as individuals, as communities, as life tethered to a single planet. Interstellar humans know a different future, one that, while not infinite, promises possibilities on timelines previously unimaginable. These are timelines commensurate with our forward-looking spirit. Our terrestrially set limits are breaking before the imminent possibility of interstellar limits just now coming into view. These new limits will continuously push old terrestrial notions into new interstellar frameworks of understanding.

It is with that in mind that I state that one of the wisest interstellar life lessons is given in Ecclesiastes 11:1. "Cast your bread upon the waters, for after many days you will find it again." It is a lesson that can be generalized to the relationship between humanity and space. So far, we cast our bread upon terrestrial waters located on the two-dimensional surface of our rock of residence, the Earth. But in the future, we will broaden our horizon along the third dimension of space and venture toward waters on other planets in the Solar system and beyond. This includes the Moon and Mars, which possess polar ice caps; Venus, where droplets may exist in the clouds; and the moons Enceladus, Europa, and Titan, where liquid water is anticipated to lie under icy surfaces. Casting probes upon these waters will enable us to search for extraterrestrial life or to establish sustainable human bases that reduce the existential risk to humanity from a single-residence catastrophe on Earth. They will also be forward bases from which humanity will go farther. Our near-future lesson will become, "Cast your craft upon interstellar space, for after millions of years you will find it again."

The wisdom encapsulated in Ecclesiastes 11:1, both in its terrestrial and interstellar articulations, is that unexpected reciprocity could be remarkably rewarding. In the long run, humanity could benefit greatly by taking the risk of venturing into interstellar space. We might discover that other technological civilizations did the same before us and that we can learn from them. Unquestionable will be our discovery that the only long run without a hard terrestrial full stop is one imagined as interstellar. This should be a vision all of humanity can unite around.

In fact, surveys conducted in 2018 and 2019 suggest that about half of all Americans are religious and that more than half of Americans believe that intelligent life exists on other planets. We do not know what percent of the religious believe in extraterrestrial intelligence, but some obviously do. And what humans have in common ought to matter more than our differences. Consider the three largest faith communities in the United States—Protestant, Catholic, and Jewish, which are always defined by their differences not their greater commonalities. However curious humans considering the possibility of extraterrestrial life and intelligence differ, they all share what confirmation of this possibility will present as: scientifically collected, scrutinized, and explained evidence.

As already exemplified by the astronauts who over years have taken up residency on the International Space Station, in the three dimensions of space, common biology binds us more tightly than competing ideologies. It takes little imagination to place yourself in a six-bedroom house (the Space Station has six sleeping quarters, two bathrooms, and a gym) orbiting Earth and grasp that the need for oxygen comes before politics. The same will hold true for our interstellar future. Consider the possibility that extraterrestrials seeded life on Earth through AI astronaut satellites pointed toward our Solar system billions and billions of years ago. To give this thought experiment a stronger grounding in probabilities, pause to note that 2.4 billion years ago Earth's atmosphere went from oxygen poor to oxygen rich due to cyanobacteria suddenly multiplying and, through photosynthesis, changing the planet's atmosphere. At that time, Mars did something akin to the opposite. It went from a planet with an atmosphere, liquid water, and the chemical foundations for forms of life similar to Earth's. About when Earth was beginning its transformation into the watery, life-rich blue marble we know from images, Mars lost atmosphere and liquids and transformed into the barren red planet we know from images and rovers.

Why these planets went on their different trajectories is unknown. Whether an extraterrestrial visit was responsible for terraforming these neighboring planets is conjecture. Less conjecture, though still highly speculative, is what would be the steps humans would need to take if they were to terraform Mars back to its condition 2.4 billion years ago. It is certain that any such undertaking would begin with scientifically informed steps to turn Mars into a planet supportive of life as we know it. Ideology and politics would be a very distant consideration after oxygen and an atmosphere.

Our interstellar future has the ability to remind us of what is really important, like liquid water mixed with salts and ammonia. So are answers to whether or not the chemistry of life as we know it is sustainable by liquid methane and ethane. The latter are common on Titan, and in 2010 the planetary scientist Sarah Hörst mixed ingredients of Titan's atmosphere, added energy such as might be delivered by the Sun's UV irradiation, and found that they could lead to the creation of the five nucleotide bases that make up DNA and RNA, the stuff of terrestrial life. Similarly, in 2013 NASA reported that complex organic chemicals could arise on Titan based on studies simulating the moon's atmosphere. And just months later, it was reported that polycyclic aromatic hydrocarbons, another piece in the puzzle of life as we know it, were detected in Titan's upper atmosphere. Given that, I wrote a paper in 2022 that notes that a hundred million years after the Big Bang the universal temperature of the cosmic radiation would have kept an object like Titan at its surface temperature of 94 degrees Kelvin irrespective of its distance from a star. The good news is that heavy elements were made in the interiors of the first stars prior to that time, allowing for our cosmic roots of early life to start when the Universe was just a percentage of its current age.

To a greater extent than ever before in human history, what happens next depends on what we choose to do. The physical reality we all share is universal, and gives us no choice but to focus on a universal narrative in making sense of it. Having made sense of it, we stand a plausible opportunity to make a world, a moon, to life-as-we-know-it's liking. Indeed, we stand a more remote opportunity to make a universe to our liking. Over a decade ago, scientists working from quantum mechanics hypothesized a creation story for everything from the laws of physics alone. Seth Shostak, a senior astronomer with the SETI Institute, summed the argument, if you "just twist time and space the right way, you might create an entirely new universe. It's not clear you could get into that universe, but you would create it." Just where lies the farthest horizon line of our interstellar future is unknown. The means of extending it, however, are. The physical reality we all share is universal, and gives us no choice but to focus on a universal narrative in making sense of it. We've already developed this narrative: the scientific method of gaining knowledge through experiments. It allows us to focus on what unites us rather than on what separates us, not just among humans but among all sentient intelligence.

THE SOUNDLESS ADVENTURE OF SENTIENT INTELLIGENCE

According to quantum mechanics, light can scatter light.

Just like any other elementary particles, the particles of light—photons—can collide with each other. And if two photons carry enough energy, their collision could produce an electron and its antiparticle—the positron. Above the energy threshold for producing electron-positron pairs, the interaction is as strong as that between an electron and a photon.

The photons emitted by the Sun could produce electron-positron pairs if they collide with other photons that are a trillion times more energetic. Interestingly, the Universe is awash with such energetic photons. They are produced in the blast waves of supernovae explosions from the collapse of massive stars, or in the relativistic jets generated by spinning black holes. The cumulative emission from all cosmic sources constitutes a background of energetic photons, so-called gamma rays, each carrying a trillion times more energy than a typical photon emitted from the surface of the Sun.

Recently, I calculated the chance of an energetic cosmic photon that grazes the Sun to be absorbed through a collision with a Solar photon. Surprisingly, this probability is as high as 7%. The extinction level of cosmic gamma rays declines by a factor of two hundred at the location of the Earth.

Altogether, it is remarkable that sunlight produces a sizable extinction of cosmic gamma rays as a result of the production of electron-positron pairs.

Nature is kind to our gamma-ray astronomers. If the Sun was ten times hotter, like massive stars, it would have blocked our view of these gamma rays from the Universe. In that case, astronomers might have not existed in the first place, because the massive star would have lasted only millions of years, consuming its nuclear fuel quickly and ending its life in a supernova explosion.

The mass budget Universe is inferred to be dominated by unknown particles that scatter more weakly than light on light. These "dark matter" particles fill up galactic space in complete obscurity, giving Rainer Maria Rilke's poem "Night" a cosmic significance. Near the conclusion of his poem, Rilke writes:

your sheer existence,
you transcender of all things, makes me so small.

Indeed, within billions of years the Sun will die, the Earth will darken—and within trillions of years, even the faint nuclear engine of dwarf stars like the closest star, Proxima Centauri, will turn off. Our ultimate cosmic destiny is to become dark matter. Unless, as Rilke encourages in the final line of his poem, we dare to bend technological selection to rewrite our civilization's destiny.

10

THE COSMIC LADDER

ARE WE ALONE? WHENEVER there is a serious, scientific inquiry into this question it inevitably attracts a lot of light and noise from the public. But in all that attention, another important question can get lost: Why are we here?

I do not believe these two questions can be disentangled. Whatever we factually discover, it will raise yet more fundamental questions. What is the meaning of our individual life, and collective lives? If a god exists, did it exist prior to the Big Bang and act as our Universe's original first-mover, or did it come into existence after the Big Bang, and if so, what is humanity's relationship to an entity possessing knowledge of the physical world and its laws vastly greater than ours but learned in a shared Universe? Whether we are unique or common creatures in the Universe has implications for how humans should treat their fellow humans. If we are the only technological civilization confronting the certainty of the end of our home planet, the eventual end of the Universe, we shoulder an awesome responsibility; if we are not, we shoulder an awesome shared responsibility.

As a *D-class* civilization we must be hopeful that a civilization higher up the cosmic ladder may aid our survival. A *B-class* civilization confronting its own challenge—how to preserve life by mastering the ability to birth a baby universe in a laboratory—could well help a *D-class* rookie that is showing precocious promise. We have nothing to lose if we adopt the attitude of such a rookie. Unique or common, our civilization stands to gain tremendously. By which I mean several thousand additional years.

We need to systematically and scientifically explore the near-Earth for extraterrestrial artifacts. We need to better and rapidly identify outlier interstellar objects, whether comet or meteor, and learn from them. We need to get more humans, more of human civilization, off our host planet. We need to invest in multiple generations of "Space Arks"—each one a better hedge against the winking out of human civilization and all terrestrial life. We need to preserve our host planet's ability to sustain life and civilization for as long as possible, so that the odds increase that answers to all of our fundamental questions are found at a pace measurable in years, at most within just a few generations.

Time is not on our side. Thankfully, optimism is.

I want you to imagine yourself somewhere strewn with trash. A refuse-filled urban dumpster, a beach prone to collecting garbage, a landfill into which the 65% of nonrecycled garbage collects, or the Great Pacific Garbage Patch swirling about in the 7.7 million mile North Pacific Subtropical Gyre. Among the garbage will be countless plastic bottles. Now try answering these questions: Which of the bottles that you see is the oldest? Which was discarded most recently?

Cosmologists have a wonderful relationship with time. When viewing space, time's arrow runs backward. The deeper we peer into the Universe's depths, the farther back in time we go. We know with precision what happened several hundred million years ago, a few billion years ago, over a dozen billion years ago, and with ever fuzzier precision what happened at the moment of the Big Bang, about 13.8 billion years ago. Humanity was given a marvelous reminder of this on Tuesday, July 12, 2022. That is when NASA released the first trove of images captured by the James Webb Space

Telescope. In them we can see the Universe 13 billion years ago, or about 800 million years after the Big Bang. They are glorious, awesome, and inspiring.

The human brain has difficulty understanding how long 13 billion years really is. To give some reference, our Solar system formed about 4.6 billion years ago as gas and dust plus gravity became the Sun and the remaining 8 planets, 181 moons, and more asteroids and comets than we've been able to count. The Cambrian Explosion, or biology's Big Bang, occurred about 550 million years ago, giving rise to complex life on Earth. Plastics were invented in 1862 and were first used to bottle soft drinks in the 1970s. With time's arrow running forward, there is a line from 13.8 billion years ago and the last plastic bottle that you discarded, I hope, in a recycling bin. Of course, hope is the tell of human sentience, because the physical Universe, past, present, and future, is agnostic in the most profound way: What you did with that empty bottle matters to the Universe not a whit. It cares about that one bottle just as little as it does about what humans do with the, by one guess, "481.6 billion bottles used worldwide in a single year."

It is a familiar human trope to imagine that there is something out there that cares about our choices, and indeed makes decisions based on our choices, whether those choices have to do with recyclable plastic or more elevated matters. There are lots and lots of examples of this, including almost all faith traditions, but let me call out a more prosaic one. In 1983, Norman Mailer wrote in *Harvard Magazine*, "I sometimes think that there is a malign force loose in the universe that is the social equivalent of cancer, and it's plastic. It infiltrates everything. It's metastasis. It gets into every single pore of productive life."

Mailer was a novelist, and so can claim some poetic license. But his analogy is deeply flawed. There are, in fact, only four fundamental forces in the Universe—strong, weak, electromagnetic, and gravitational—and none of them are malign. Additionally, cancer, or the uncontrolled growth and spread of damaged or abnormal body cells, is a biological quirk of natural selection that existed long before its first diagnosis in humans. Evidence of cancer has been found in dinosaur fossils, and in 1932 the paleoanthropologist Louis Leakey identified a malignant tumor in Australopithecus. While the spread of plastics seems to have a correlation with human diseases, the threat is, like these plastics, entirely human-made.

Only sentience might care about other sentience. And it is on that hope that humanity must hang all of its hopes.

We might do better to think of the planetary expanse of plastic garbage and Webb's deep images of the early Universe as akin to a Magic Eye poster, where if you just shift your focus a bit hidden things pop into view. We are approaching a tipping point on the seesaw of natural selection and technological selection. At the largest scale, knowing what happened 13 billion years ago helps us anticipate what is likely to happen 13 billion years in the future. And at the smaller scale, knowing that within half a century we've generated all that poorly biodegradable trash helps us anticipate what is likely to happen fifty years in the future. And both should tell us what we must now do.

The next fifty years, and the next billion, will depend increasingly on technological selection displacing natural selection. That is one of the hidden messages in Webb's images. If we are to persevere, thinking of how technology selects for progress over the next billion years is humanity's challenge, opportunity, and sacred trust.

Yet that message about technological selection is not the only, and certainly not the first, thing we see in Webb's images. What often overwhelms us on first glimpse is beauty, and with it a sense of awe. There, in reds and blues, swirls and bursts of light, is our origin story. And not just ours, but the origin story of everything we see, touch, feel, and through the gift of sentience turn our conscious thoughts to.

It is natural to focus on the luminous islands of starlight in these images. However, these represent a small fraction of what is in the frame, literally the tail of the dog of matter. The dark regions between them are, we are nearly certain, full of dark matter that dominates the mass budget and the resulting gravity. Therefore, it is dictating how these luminous cores move. But more accurately, the dark regions represent our ignorance, since we still do not know what the dark matter is. The second message delivered by Webb's deep, ancient images is the limited extent of humanity's current knowledge.

We must explore what we do not know, because it often carries more weight than what we know. Only by exploring the unknown will humanity be able to become aware of the full scope of its existential risks and survive them. And unlike the Universe and all the non-sentient matter it contains, humans have proven themselves consistently concerned with continued sentient existence, even when concern and sentience is prejudicially defined.

Human history, the measure of human civilization, is shot through with passionate beliefs. Time's passage, a measure of the Universe, isn't. But whereas the Universe and its feature of relativistic time need pay no attention to human civilization, we must pay attention to both if we are to continue. When a 102-year-old woman was asked why she continues to smoke rather than quit based on the advice of her doctors, she replied: "All the doctors who gave me that advice are dead by now." This is a reminder that in all existence determined by the survival of the so-called fittest, time is the ultimate arbitrator.

For millennia, natural selection has trained us to think short-term. When your options are fight, flight, or freeze, those members of a species most ferocious, fleetest of foot, or best camouflaged are the ones who will live to procreate another day. In this fact is an answer to a question the Israeli writer Dror Burstein emailed me on noting the front page of newspapers placing Webb's deep images of galaxies alongside stories about wars and corrupt politics. "Why," Burstein wondered, "do these items attract the same level of attention?" Because heretofore, we have almost always co-opted technological selection to aid natural selection, with our politicians' presentations of facts (very often flawed, false, or incomplete) preceding state-level violence inflicted on weaker neighbors. That is our civilization's current perfecting of fight, flight, or freeze. Wars are always stories of technologies used to advance such short-term objectives.

When we consider the biggest item on our agenda, survival of humanity, obviously we must do better. Our predicament viewed only on humanity's small scale can seem depressing. Viewed, however, on the large scale, we should be optimistic. The vantage point of 13 billion years suggests strongly that there is an extraterrestrial helping hand awaiting our sufficient curiosity.

Enrico Fermi's 1950 lunchtime question, "Where is everybody?" was, of course, a statistical query. Given the amount of useful data we have about extraterrestrials, it can be reframed today as, "What can we infer about the existence or extent of ETI?" When Fermi posed the question, that data set was zero and all inferences valueless. Since then, our collection of useful data has increased marginally, but not negligibly. In just the last five years, our data have allowed us to increase the number of confirmed interstellar visitors to the Solar system from zero to four. Three of them, 'Oumuamua, IM1, and IM2, demonstrated sufficient outlier properties compared to other comets and meteors to spark heightened curiosity, and the plausibility of extraterrestrial manufacture. And in a matter of years, with the Galileo Project, NASA's unidentified aerial phenomena study, and most significantly the Legacy Survey of Space and Time (LSST) on the Vera C. Rubin Observatory going live in 2023, we will gather more, and ever more useful, data.

From the Kepler satellite data on exoplanets we know that a substantial fraction of Sun-like stars have a planet the size of the Earth roughly at the same separation. Since most of these stars formed billions of years before the Sun, the dice of intelligent life was rolled tens of billions of times within our own Milky Way galaxy alone. It is very likely that Albert Einstein was not the smartest scientist that has lived in the Milky Way since the Big Bang. Rather, it is likely that smarter scientists resided on an exoplanet around another star and the civilization that benefitted from their wisdom launched probes into space a billion years ago. If these probes were self-replicating, then their descendants can be all around us by now and all we need to do is search for them. Most of these probes might have been too small or fast to be identified by past astronomical surveys of the sky.

There is also so much more we can do. For just one example, if we avoid the logical fallacy of Carl Sagan's belief that some claims are extraordinary and so require extraordinary evidence and adhere to the scientific method's straightforward expectation that claims require evidence, we can take fuller advantage of the hypothesis that considers 'Oumuamua an alien craft, and act accordingly.

In April 2022, my colleague Martin Elvis published a paper in *The International Journal of Astrobiology* considering the research programs that could arise from the hypothesis that 'Oumuamua was an extraterrestrial artifact, "an alien craft dominated by a solar sail" that was under either controlled or uncontrolled flight. Approached this way, an extraterrestrial 'Oumuamua presents questions of known physics and engineering challenges. The limited number of "possible scenarios that flow from the assumption that 'Oumuamua is an alien craft" allows for more rigorous research, which would have applications "to any future objects for which" an extraterrestrial hypothesis "is raised."

The implications are not just for what we might do next time a UAP enters our Solar system. Even without having photographed or captured 'Oumuamua, the scientific method allows for theoretical reverse engineering. This is further encouraged by Elvis bounding his assumptions to an understanding that the craft's designers "worked with the same physics we have, the same fundamental constraints on materials that we have, and design constraints that are familiar to us." What they might have built, we might build. What became of what they built could represent the future for crafts we might build.

Precisely because we failed to image or capture 'Oumuamua, we have mostly questions that, were we to act on them, could prove productive.

Was it a derelict vessel similar to cosmic trash? That suggests a civilization or civilizations whose profligacy with stuff is reminiscent of humanity's profligacy with trash and pollutants. Was it a guided message in a bottle, coy or otherwise? By one set of assumptions, this suggests a civilization capable of thinking, and acting, on scientific projects whose duration is measurable in millions of years. An extraterrestrial 'Oumuamua directed toward the Solar system for our potential benefit would have been launched before any evidence of human sentience would have been detectable. This would make it a civilization's modest, hedged-bet nudge of "you are not alone" very similar to humanity's two Voyager craft. This also suggests a civilization able and willing to send out millions, perhaps billions, of such hedged bets, selecting for the most promising exoplanets.

If sent with intention, where did it originate from? Given its hypothesized solar sail shape, did it originate from a star or might artificial means—such as a radiation beam—been used to keep it in the Local Standard of Rest? If the latter, might it have been more akin to a buoy our Solar system ran into than a craft sent in our direction? Given different points of origin, how many course corrections would have been necessary for 'Oumuamua to attain its unique trajectory through the Solar system? What if 'Oumuamua came not from the star system where it was built, but rather from a way station chosen for its gravity assistance?

And there are the nontrivial engineering considerations. An alien-constructed 'Oumuamua would mean a craft engineered to persist around twenty times longer "than the oldest human-built structures on Earth" and many thousands of times "longer than the longest-lived spacecraft" humanity has ever launched. And yet, the ongoing construction of the Clock of the Long Now, once built expected to last ten thousand years, and the research undertaken for the one-hundred-year spaceship indicate a surmountable challenge. This is especially true as the stability of human-made materials increases with our technology. Diamond sheets, better able to survive encounters with interstellar dust, are currently beyond our technological abilities, but theoretically possible. As are thin superconducting magnets to create an artificial shield to protect the sail from cosmic rays.

The conclusion of Elvis's paper is that whether or not 'Oumuamua was a message in a bottle informing us that we are not alone, or jetsam tossed with all the consideration of a plastic bottle out a car window, humanity could take even one valuable data point—an ambiguous interstellar object—and bend the course of our civilization upward. "Overall," Elvis concludes, "these considerations show that a broad and well-defined research program can be built around the hypothesis that 'Oumuamua is an alien craft. More generally, the considerations presented here can also be applied to other interstellar visitors, as well as to general discussions of interstellar travel."

PUSHING THE HAWKING LIMIT

Achieving goals against all odds makes life worth living and exemplifies the resilience of a sentient spirit. For me, the most inspiring example is Stephen Hawking, who, despite being paralyzed by a motor neuron disease and for much of his adult life unable to speak, became one of the leading physicists of his generation. For half a century, he was dependent on a wheelchair, initially activated by a hand switch and eventually, when he could not move his fingers, by a single cheek muscle, facial expressions, and eyebrow movements.

This was his condition in 2006, when Hawking stated that his greatest unfulfilled desire was to travel to space. The following year, he flew aboard a reduced gravity aircraft off the Florida coast to experience weightlessness. A decade later, he joined me on a stage in New York City for the public announcement of the first funded initiative for interstellar travel. This was followed by a visit to Harvard University for the inauguration of the Black Hole Initiative for which I served as the founding director. My most memorable experience with him was when he joined my family and friends for a Passover dinner. The best story I heard about Hawking from that trip was when he told his caretakers one evening, "I am bored. Let's go to the hotel bar and have some fun."

Hawking died in 2018, and he continues to inspire us to ever advance the Hawking Limit, or the extreme condition when most of the human body loses functionality but our sentient intelligence remains productive, even joyous. At this limit, we can ask, what establishes human identity? It is certainly not a fully functioning body, although it definitely requires a functioning human brain, or its AI equivalent. French enlightenment philosopher René Descartes famously declared, "I think, therefore I am," which applies to intelligent sentience, not just to humans.

For Descartes, sentient awareness was a starting point for a philosophical exploration of ontology. For human civilization, sentient intelligence was the starting point for the long, slow application of technology to the scales of natural selection. From the warmth gained by turning animal hides into clothes to robotic factories helping in the production of automobiles with driver-assist functionality, the application of evidence-based technologies to human civilization's problems is constant, even their application to the problems created by earlier technologies.

No branch of human civilization has reached its level of competence without help from others. Throughout our civilization's history, humans have poached technological insights from their fellow humans. Knowledge contained in ancient Greek texts traveled to the Arabic civilizations and then returned to Europe. Technologies like silk and gunpowder were carried from China across long trade routes to the West. Shipping containers and globalization have knitted together all the nations of the planet. And then, of course, there is the near-instantaneous transmission of information on the World Wide Web. Whether extended in friendship or not, technological assistance across civilizations is a human constant. In our near future, such assistance will come by way of AI, if we build it, and ETC, if we seek it.

NASA's Perseverance rover is not sentient. The six-wheeled motorized robot, which touched down on the Red Planet on February 18, 2021, does have an auto-drive function, AutoNav, which allows it to find optimal paths around rocks and obstacles. But otherwise, its functions are dependent on an Earth-bound team of scientists in Pasadena, California. Wearing 3D goggles, they issue commands that take twenty minutes to reach the rover; these drivers—there have been approximately a dozen of them—experience something akin to "playing a virtual reality video game." One goal of the game is to seek evidence of microbial life on Mars.

At the time that Perseverance was conceived and built, Earth-based control of it was the only reasonable option. Had an astronaut, or a dozen of them, been sent to pilot Perseverance on Mars, about a year into their Martian stay we'd have to start to think of ways to bring them home lest some start to sicken from UV radiation. We should celebrate what Perseverance points toward: the proliferation of human-controlled avatars on the planets of the Solar system, and AI astronauts in interstellar space. At minimum, such monuments to humanity can exist long after a catastrophe wipes out all the statuary in all the museums on Earth.

The advent of AI astronauts is quickly coming. A twenty-minute delay driving Perseverance is manageable, but the delays required to pilot a similar technology on Alpha Centauri B, four light-years away, are absurd. Our ability to promote humanity, and our civilization, in space depends on the phase transition from sentience that is housed exclusively in terrestrial biology to, as we push the Hawking Limit, sentient AI. Whatever a craft's means of propulsion, technological equipment will be more resilient than living organisms when traversing interstellar distances given the harsh conditions, which include bombardment by energetic cosmic rays and dust particles as well as harmful radiation in the UV and X-ray bands. Such craft must possess their own intelligence and ability to operate on their own initiatives, following guidelines received on Earth. This will be different, but by a matter of degree not of kind, from the way our children explore the world with ever-increasing confidence as they grow up.

It is time to recognize that the forces that will shape our future will be much different from those that governed our past. Natural selection, which dictated life for billions of years on Earth, must be replaced by technological selection for billions of years in space. We cannot afford another billion years of gradual biological evolution on Earth, since the Sun will boil off all our planet's oceans within that time frame. We are nearing the end of our cosmic journey on this host planet, and frankly going around the same star, which we call our Sun, for so long is rather boring.

From now on, science and technology will determine whether life-as-we-know-it will survive and for how long. To promote long-term survival, we must shift attention from short-term risks and focus on what matters in the long run. When contemplating survival plans, we must remember a quote of William of Occam's: "Entities are not to be multiplied beyond necessity." It is a lesson writ-hard into the logics of natural selection. From eukaryote cells to us, nature indiscriminately acts on the fact that there is energy efficiency in pruning possibilities. Humanity has, of course, throughout its history adapted that lesson, though often with hubristic biases and cruelty. Technological selection, by application of the scientific method, can do better.

For optimists, humanity's route up the rungs of the cosmic ladder of civilizations is by one of two paths. We manage it with the assistance of an extraterrestrial technological civilization or its artifacts. Or we manage it entirely on our own. If the latter is the only path we choose, or discover, optimism dims.

A logical bookend to Clyde Tombaugh's ashes traveling in space is the bunker built beneath the White House in Washington, DC. Along with a number of Cold War–era fortifications in the capital intended to allow a percent of our politicians to survive a nuclear weapons strike, a secure bunker exists to protect the executive branch. By just an impressionistic, unscientific consideration, this is a strategy not likely to preserve the best that humanity's DNA has to offer. The same can be said if wealth alone determines who surmounts the risks and dangers of a settlement on Mars, or a berth on a solitary spaceark built to sustain select pairs of life.

Space travel should not be guided by commercial benefits because there is no way to cash in on an exit from the Solar system. Instead, interstellar journeys should reflect humanity's spiritual quest for exploring the unknown, getting a better sense on how it came to exist, and teasing out our purpose.

It is possible that complex life was seeded on Earth by another civilization that reached the same conclusion about its long-term survival as is increasingly dawning on a rising percentage of Earth-bound humans. In that case we have interstellar relatives to search for. Or seed.

As John Chapman knew, if you want apples, plant apple trees. Known to history and story time as Johnny Appleseed, Chapman is credited with spreading fruit trees from New England to western Ohio. He was vastly aided by the fact that apple trees seed their own replications. A question the physicist John von Neumann took up was, Could machines do the same? What biology does so successfully by natural selection, could humanity do by technology?

Simple self-replicating models have been built since the mid-1950s, most memorably by Lionel Penrose and his son, the physicist Roger Penrose, working with plywood (a video of their effort is viewable on YouTube). For efficiency, von Neumann elected to work by computer simulations of cellular automata. A result has been self-repairing silicon chips and autocatalyzing molecules. Von Neumann theorized a universal constructor capable of creating, and re-creating, any pattern. Subsequent work by a string of scientists has further refined the problem, and its solutions, illuminating "the underlying principles of replication and, by doing so, [inspiring] more concrete efforts" at building a machine capable of duplication at a rate somewhere between a sequoia and a weed.

Self-replicating machines. Sentient computers. AI astronauts. CubeSat arks sparking new life by 3D printer technology on distant planets with a different sun, or suns. If human civilization's fifty-year horizon seems fraught with risk and uncertainty, danger and an ever-mounting mound of trash, then our billion-year horizon can seem so utopian in its solutions that they promise human civilization endures but not as anything recognizably human. The lowest rungs of the civilizations ladder seem identifiable enough, but also concerning. If a *C-class* civilization is one that is not degrading its host planet's ability to be habitable, we're a rung beneath it. A *B-class* civilization, one able to exist in an environment independent of its host star, is a significant leap up from *C-class*, but one in which bipedal humans, at some Hawking Limit, are visible. But it is at the *A-class*, or a civilization capable of recreating the cosmic conditions that gave rise to existence, namely a civilization capable of creating a baby universe in a laboratory, where even the furthest edge of the Hawking Limit seems to vanish.

As questions for science to answer, these can seem impossibly distant from our current considerations. Overcoming that sense of distance is achievable, I believe, by a familiar, well-traveled path. Herein lies, I find, a gift of scientific spirituality as Rabbi Rob Dobrusin touched on when he allowed that the search for extraterrestrials was not only not at odds with faith, but was a search that could be "extraordinarily meaningful on a spiritual level." For millennia, humans have not only sustained faith in meaning and purpose provided by religious covenants with God, but they have imbued that faith in their families and communities and sustained institutions and practices to safeguard and advance their faith. The search for extraterrestrial near-Earth artifacts, the decision to support spaceark programs to better safeguard terrestrial life and human civilization, the determination to ascend to the next, and the next, rung of the cosmic ladder of civilizations: these are the practices of a civilization that holds to the tenet that preservation of life and the possibility of civilization's persistence is an encompassing meaning for humanity.

Discoveries that await our immediate tomorrows sit within a timeline that stretches back to the Big Bang, but also, potentially, stretch out along a timeline that ends in an *A-class* civilization's birth of a new universe.

Achieving the distinction of an *A-class* civilization is nontrivial by the measures of physics as-we-know-it. But it is achievable by theorizing quantum effects that allow a small bubble of false vacuum to tunnel into a larger bubble "of the same mass, which would then classically evolve to become a new universe." At the outer edge of a civilization's use of technological selection are answers to where we came from, what is our purpose, what is the meaning of our sentient intelligence. Perhaps such intelligence arises for the purpose of attaining the last rung of civilization's ladder. Meaning lies in not just whether but how this is achieved. Our purpose is continuity: sentient intelligence in complete mastery of physical laws and matter created sentient intelligence. These are tasks entrusted to all sentient intelligence in the Universe, and that is reason to anticipate and seek out other sentient intelligences' help while ever preparing ourselves to offer our own.

A decade ago, I wrote two textbooks, one titled *How Did the First Stars and Galaxies Form?*, and the second coauthored with my former graduate student Steve Furlanetto, titled *The First Galaxies in the Universe*. Both books describe theoretical expectations for what the Webb telescope might find in the context of the scientific story of genesis, "Let there be light." More recently, I coauthored a textbook with my former postdoc Manasvi Lingam, titled *Life in the Cosmos*. I would, of course, be thrilled if the forecasts in these textbooks were confirmed by future Webb data. Even better, though, would be the civilization-wide thrill if Webb's data surprised us with new discoveries that were never anticipated.

Time's backward-pointing arrow bull's-eyes the Big Bang. And among cosmology's biggest mysteries is what happened before it. A century ago, Albert Einstein found the idea of a beginning in time so philosophically unsatisfying that he searched for steady-state alternatives. There are none.

By now there is a variety of conjectures for our cosmic origins in the scientific literature, including the ideas that our Universe emerged from a vacuum fluctuation or that it is cyclic with repeated periods of contraction and expansion or that it was selected by the anthropic principle out of the string-theory landscape of the multiverse, where "everything that can happen will happen an infinite number of times," or that the Universe emerged out of the collapse of matter in the interior of a black hole. What these ideas largely or entirely exclude is purpose and meaning. Or, perhaps more accurately, it is what they evade.

The possibility that our Universe was created in the laboratory of an advanced technological civilization provides humanity both. Since our Universe has a flat geometry with a zero net energy, an advanced civilization could have developed a technology that created a baby universe through quantum tunneling out of nothing. This possible origin story unifies the religious notion of a creator with the secular notion of quantum gravity. Humanity does not now possess a predictive theory that combines the two pillars of modern physics: quantum mechanics and gravity. But a more advanced civilization might have mastered the technology of creating baby universes and accomplished this feat. If that happened, then not only could it account for the origin of our Universe but it would also suggest that a universe like our own—which hosts an advanced technological civilization that gives birth to a new flat universe—is like a biological system that maintains the longevity of its genetic material through multiple generations.

If so, our Universe was not selected for us to exist in it, as suggested by conventional anthropic reasoning, but rather it was selected such that it would give rise to civilizations that are much more advanced than we currently are. Those "smarter kids on our cosmic block," the cumulative talent of the cumulative sentient intelligence in the Universe, which are capable of developing the technology needed to produce baby universes, are the drivers of the cosmic Darwinian-technological selection process.

Genesis 3:19 states, "By the sweat of your brow you will eat your food until you return to the ground, since from it you were taken; for dust you are and to dust you will return." The sentient intelligences that stand atop the cosmic ladder of civilizations can add, “and return, and return, and return.”

Any civilization that has existed over the past 13.8 billion years, or will come into existence in the billions of years remaining to the Universe, might follow the trajectory of natural selection on their host planet to technological selection among the stars. If so, this guarantees that all such civilizations will have confronted the lower rungs of the cosmic ladder. They will have managed not to degrade their planet before the laws of physics do. They will have managed to replicate habitable conditions off their host planet. They will confront the same fundamental questions of existence and spy the possibility of the ladder’s highest rung. The most meaningful pursuit of any sentient intelligence is to scientifically understand the details of the cosmic circumstances that led to their, and our, existence.

Inevitably, humans will add the last plastic bottle to the terrestrial collection of plastic trash. There is, of course, a last time for everything. There will be a last time each of us draws breath, a last time the Earth orbits the Sun, a last time the galaxies in our ever-expanding Universe are visible from any other galaxy, and a last time when wondering if humanity is alone in the Universe remains of interest. We may or may not discover the answer before we cease. But perhaps we will participate in a cycle, virtuous or not, of sentient intelligence rising to the level of birthing more sentient intelligence, some of which eventually arise to the point of birthing another universe.

The catechism for science-directed spirituality in pursuit of the most fundamental questions of existence is brief: ever more useful data rigorously applied to scientific inquiry with egoless transparency of discoveries among other sentient intelligence. In place of prayer there is the scientific method.

CONCLUSION

YOU ARE NOT ALONE.

Every human knows that somewhere there are, or were, parents; are, or were, grandparents and relatives; are, or were, other humans. When we apply our reasoning intellect to how unique we individually are, we grasp an origin story that links us back through the thousands and millions of years to eukaryotic fossils. On Earth, to be unique, incomparable, solitary is not the condition of either humanity or any form of life.

Are we alone in the *Universe*?

We don't know. We have barely bothered to look; we have barely bothered to apply our reasoning intellect to the problem. Our generation is, however, finally preparing to do this long-overdue scientific work. Everything could change, and should change, as a consequence.

For thousands of years humans believed the Sun—chased by wolves, pulled by a chariot—went away at night and the Moon came out. For less than a thousand years, an ever-increasing number of humans understood that darkness signaled we stood on the side of the Earth facing away from the Sun, and the Moon was illuminated by the solar lamppost. Earth, our blue marble, is never alone, for even with sunlight blocked by Earth we can infer the Sun by looking at the Moon.

So, when you step out at night you can observe the Sun by observing the Moon—but there are limits to what you can understand even so. Sunlight reflected off the Moon can never fully deliver us a perfect rendering of the Sun's properties. As with any reflection, some data must necessarily be lost depending on the qualities and limitations of the reflecting surface. Just as grandchildren tell us something about their grandparents, it will always pale to what we can understand by meeting the grandparents.

Consider another example. We infer that dark matter exists from its gravitational influence on ordinary matter in the form of stars, gas, and light. The motion of stars and gas in galaxies, the pressure of hot gas in clusters of galaxies, and the effect of gravitational lensing by all cosmic structures imply dark matter. Similarly, the expansion rate of the Universe suggests that the cosmic mass budget contains five times more dark matter than ordinary matter. Still, while inferences are important, especially when they are all we can access, they cannot replace immediate interaction with the object under study. We have much more to learn.

A useful morning salutation, to follow or even proceed "How are you?" would be, "Our scientific knowledge is spotty, like islands in an ocean of ignorance." The best visual illustration of this state of affairs is the first image from the James Webb Space Telescope, which in a kaleidoscope of blues, oranges, and reds shows a cluster of galaxies that gravitationally lenses light from even more distant galaxies, and in the process magnifies just-born dwarf galaxies dating all the way back to 13 billion years ago. Our eyes and imaginations are drawn to the colors, and not to the vast dark that surrounds them. It is a signature of human-centrism to presume that most of the interesting action in the Universe involves the minority of matter that we see, just because we are familiar with it. Such a bias echoes the mindset of children who imagine that there is nothing exciting outside their home. Until, of course, they are introduced to their neighborhood, neighbors, the toys of distant playgrounds, and playdates.

To find new things, to enlarge our islands of knowledge, to discover or construct land bridges is the work of science and technology. For decades, in hopes of producing dark matter, we've smashed familiar particles against each other, first in cyclotrons and then in synchrotrons. So far, all of our attempts to produce dark matter in particle accelerators have failed. At the same time, the history of relativity and quantum mechanics suggests that nature is more imaginative than we are.

More concerning than anything not terrestrial is the tenacious grip of humanity's terrestrial history. When the United States Air Force shot down a Chinese balloon, presumptively sent over America for reasons of espionage, many in the media declared the two superpowers had entered a new era of contest, dragging the rest of the world with it. The opposite is true. Terrestrial nations spying on each other with the most technologically advanced tools is an ancient practice. Briefly imagine a history in which, alongside SETI's founding in 1984, a Galileo Project was founded and funded so that its UAP observatories were ubiquitous across the globe. Decades of peer-reviewed data about what is up in Earth's skies and atmosphere would have, at minimum, ended any effort to use high-altitude surveillance technology for spying. And perhaps, just perhaps, decades of data would have winnowed down UAP to just a handful, and these would have then become the focus of global scientific scrutiny.

An interstellar nudge that jump-starts our imagination could advance human knowledge, and civilization, rapidly, allowing us to accomplish in one generation what might otherwise take several. Otherwise, we are bound to what humanity, by its own initiatives, manages to understand.

In the same way that the nearby Moon educates us about the faraway Sun, the laws of physics that were uncovered in laboratories on Earth appear to reflect the Universe at large. This is not a tautology. We could have observed laboratory-measured parameters, like the speed of light, the electron charge, or Newton's constant, to have different values at different cosmic locations or times. Instead, we discovered the opposite to be true. They are constant to better than a few percent all the way back to four hundred thousand years after the Big Bang, when the Universe first became transparent.

Finding that the physical laws uncovered in our laboratories reflect the behavior of the Universe at large, like in Plato's allegory of the cave, raises the question of whether sharing fundamental laws implies also sharing phenomena. More concretely, we might ask, Is the chemistry of life on Earth replicated on rocky exoplanets with similar surface temperature, water content, and a thin atmosphere on top?

By studying Earth, we may not acquire a good understanding of what lies afar. For example, the composition of the Sun is different from that of Earth, as Cecilia Payne-Gaposchkin first realized through her PhD work in 1925. This was contrary to prevailing presumptions governing at the time, and the then director of the Princeton University Observatory, Henry Norris Russell, wrote to her that her findings were "clearly impossible," only to realize that she was right through his own work conducted four years later.

This brings us full circle to the point that studying the sunlight reflected off the Moon would not reveal the nature of the Sun as a nuclear reactor confined by gravity. For that purpose, we need to study the Sun directly during daytime, as Payne-Gaposchkin did.

In July 2017, I visited the Daniel K. Inouye Solar Telescope (DKIST), when its four-meter telescope was being assembled on Haleakalā in Maui, Hawaii. During the tour, the local astronomers were arguing that they will not be getting enough light from the Sun to study it in desirable detail, a debate that echoed the endless nature of the pursuit for scientific knowledge: Do we have what we need to discover the data that will help us understand what we currently do not? Within a few years, DKIST delivered amazing high-resolution images of convective cell-like structures, the size of Texas, on the surface of the Sun. The result: the frontier of the debate was moved further out to the next instrument needed to gather yet more data to help expand our knowledge yet again.

Remote observing is the first step toward scientific knowledge. We get limited information about our neighborhood by looking through our windows. A better approach is to venture into the neighborhood, in the spirit of NASA's Parker Solar Probe. By plunging repeatedly into the Solar corona since its launch in 2018, the probe is unraveling new details about the Sun's magnetic and energetic particle phenomena.

Human sentient intelligence has for centuries and centuries plunged again and again into the details of existence. The more we have done so, and the more recent the plunges, the more confirmation we receive that the Universe is far more imaginative than we are. Consider that the uncertainty principle of quantum mechanics allows that a car can in principle pass through a brick wall intact. The chances of it happening are negligibly small, though. Elementary particles, much smaller than a car, have a higher likelihood for quantum tunneling through a barrier. This is because they are not localized as well as a massive object, and the wave function that characterizes the probability distribution of their uncertain location has an extended tail that could surpass the barrier.

Such tunneling occurs during the nuclear fusion in stellar interiors, and a result is heavy elements, like oxygen and carbon, that are essential for life. Such tunneling might also explain how dark matter particles, if postulated to be light enough, escape the gravitational potential wells that bind them to dwarf galaxies within the halo of the Milky Way. This would relieve the tension between theory and observation, and settle one of the mysteries of the still invisible dark matter. It would also allow quantum mechanics, constructed to explain the behavior of the smallest systems we know, such as elementary particles bound in atoms, to be used to study some of the largest bound systems we know, galaxies. It isn't just that the natural world is weirder than we know, it is what the natural world is that awaits our sufficient attention.

According to the ancient Greek legend, Sisyphus was condemned by the gods for eternity to repeatedly roll a boulder up a hill only to have it roll down again once he got it near the top. Albert Camus used this story as a metaphor for our persistent struggle against the essential absurdity of life. Perhaps Camus approached the problem from the wrong end of the trench. Approaching it from philosophy, he found himself questioning suicide. But, if approached from the known science of physics, the natural world at least presents a solution to Sisyphus's absurdity. In enabling tunneling, quantum mechanics removes the need for his existential struggle. It asserts that if we just keep waiting with patience, the boulder will end up on the other side of the hill by itself.

And with that problem solved by nature, Sisyphus could take up a challenge his labors could more measurably advance.

ILLUMINATING A DARK UNIVERSE

Everything humanity has learned about life as we know it, and everything humanity has learned about the Universe's extent, laws, properties, and matter as we know it, tells us we are not alone. That over all that time and all that space the circumstances that led to sentient intelligence arising on Earth occurred only once is extremely improbable. If only one extraterrestrial, technologically advanced civilization survived to create artificially intelligent, self-replicating spacearks, the probability of there existing a statistically significant amount of extraterrestrial artifacts near Earth arises. If there were several such civilizations, that probability increases. That humanity is likely within a generation of sending its own AI CubeSats equipped with 3D printers out into space increases the probability that many more than just two similarly capable civilizations existed over the nine or so billion years before life emerged on Earth.

That two of the three interstellar objects humanity has identified over the past decade exhibited outlier properties when compared to data of stellar objects is an intriguing admission of human ignorance. There is something here we know far too little about. And there is something here that is begging our sufficient curiosity to gather the data to narrow the extent of our ignorance.

Plunging into interstellar space would require a major investment of time. Before pursuing that, we should examine the objects that enter our Solar system from interstellar space. And while pursuing those objects, we should examine the interstellar meteors on Earth. It is a statistical certainty that there are more than two, but an excellent place to start is with the two, IM1 and IM2, we know of.

On April 18, 1955, the day Albert Einstein died, illustrator Herb Block drew a famous cartoon of the Earth labeled with the plaque ALBERT EINSTEIN LIVED HERE. If, perhaps by a not-implausible gift of the Hawking Limit, Einstein were alive today at age 143, he would be thrilled to see many of the phenomena he theoretically forecasted a century ago being observed in the real Universe. These include the cosmological constant, black holes, gravitational waves, and gravitational lensing. But I can also imagine him stating, "It is high time for us to find something really new about spacetime and gravity that I had not anticipated."

Given that Einstein is not around anymore to guide our future scientific endeavors, we should seek advice from interstellar space. Without such help, we are reduced to searches bounded by terrestrial limits of knowledge, imagination, technology, and our biological and cultural frailties. Like particle accelerators, we can, and likely will, keep slamming things together at ever-increasing energy in hopes of a different result.

My hope lies in the unexpected insights to be garnered through the new Webb telescope, the Vera C. Rubin Observatory, and the new telescopes, deepwater sleds, and ambitions of the Galileo Project. My optimism arises from the probability of our discovering near-Earth extraterrestrial artifacts that immediately reframe humanity's understanding of its place and purpose in the Universe, making ever more plain that meaning lies in the difficult but attainable work of ascending the cosmic ladder of civilizations.

Then I recall the vast majority of the Universe that I cannot see and about which humanity is profoundly and almost entirely ignorant and my optimism increases. So far, all of our attempts to produce dark matter in particle accelerators by smashing familiar particles with high energies have failed. Why?

The simplest assumption to make is that dark matter particles do not scatter off each other, just as they do not scatter off ordinary matter. They simply do not interact with matter, ordinary or dark. This assumption, however, leads to tensions between current theory and current observations of dwarf galaxies. What may help resolve the tensions are—no surprise—more data. Observing distant dwarf galaxies, out of which our Milky Way was assembled at early cosmic times, would allow us to probe the nature of dark matter. Such data will soon be obtained: extending the Webb telescope's exposure time to three weeks would allow it to detect dwarf galaxies that are six times fainter than those seen in its preliminary deep image.

Perhaps we will learn that the challenges to noninteracting dark matter are alleviated by discovering that dark matter is self-interacting. Over a decade ago, the astrophysicist Neal Weiner and I hypothesized that dark matter was, like electrons and protons, electrically charged particles that moved at velocities far below that of ordinary matter. A consequence would be noninteraction with the matter we see. Dark matter would only interact with dark matter. It is a hypothesis that explains some observable data, namely visible differences between the behavior of clusters of galaxies and dwarf galaxy cores. It is also a hypothesis that would call for a new physics beyond the standard model. For now, it is one hypothesis competing among others. When pondering our interstellar future, however, it is a hypothesis that suggests a fascinating set of possibilities.

Self-interaction allows for the possibilities that there are "dark atoms" that emit "dark radiation" and condense into "dark stars" and "dark planets" that support "dark chemistry" that leads to "life-as-we-do-not-know-it." In such a case, an answer to Fermi's question, "Where is everybody?" might be, "We are blind to most of them." Perhaps in such a dark universe, which enjoys at least 80% more dark matter than our "ordinary" matter, there are yet more technologically advanced civilizations that have mastered the physics necessary to observe us. And if anyone out there is using dark matter or dark energy to fuel their spacecraft, we will see nothing coming out of their engine's exhaust. What we can hope, and with some confidence, is that they are animated by the same spirit of scientific discovery and progress as animates the most hopeful elements of human civilization.

The scientific method is guided by data. Favoring evidence over prejudice has two manifestations, one practiced routinely and the other mostly ignored. The first manifestation uses the guillotine of data to eliminate possibilities that do not describe reality. The less practiced principle is to regard evidence that deviates from theoretical expectations as an opportunity to learn something new rather than as a threat to expertise that rests on past knowledge. The spirituality of the scientific enterprise rests overwhelmingly in this second, less practiced principle.

Here lies a sacred aspect of the Copernican revolution. Not only are we not at the center of the cosmic stage, not only did we come late to the stage, but life as we know it among matter as we know it does not even represent most of the stuff that is presented on that stage. We are more Rosencrantz and Guildenstern than Hamlet and Horatio, and until we figure out what this dark stuff is, we cannot claim to know what the cosmic play is about.

And yet we are close, tantalizingly close, to grasping the next rung up the cosmic ladder of civilizations. To the moment when technological selection obviates natural selection. To an empirically proven theory knitting gravity with quantum physics. To new and far, far faster means of propulsion. To sentient artificial intelligence, and the pushed outer possibilities of the Hawking Limit.

All that we currently know about the Universe and our place in it suggests a scientific covenant. The meaning of our existence rests in our science-led persistence as sentient intelligence, as a civilization. Therein lies not only our purpose, but the shared purpose of all sentient intelligence throughout the Universe, dark and ordinary. For at the highest rung of the cosmic ladder of civilizations, there is no dark, there is no ordinary, there is only fulfillment of a sacred trust: continuance.

ACKNOWLEDGMENTS

I am most grateful to my wife and closest friend, Dr. Ofrit Liviatan, and to our daughters, Klil and Lotem, whose unconditional love and richness of talents inspired the forward-looking perspective of this book. If I ever embark on an interstellar trip, they will be the first to be offered a seat next to me.

I am also grateful to my many research collaborators and in particular my brilliant student Amir Siraj, with whom I coauthored three dozen papers in the three years preceding this book.

In addition, my thanks go to all the visionary donors who generously supported my research agenda in recent years, starting with Eugene Jhong in May 2021. It has been a special privilege and pleasure for me to partner with Dr. Frank Laukien in founding the Galileo Project and the Copernicus Space Corporation, which blossomed into hubs of exceptional talent and innovative research. My gratitude also goes to Charles Hoskinson for funding the expedition to retrieve fragments of the first interstellar meteor from the Pacific Ocean floor.

This book would have never come to fruition without the editorial team members. In particular, I am grateful to my literary agent, Leslie Meredith, for her guidance and support; to the editors Matt Harper, Georgina Laycock, and Alex Littlefield for their exceptional advice; and to Thomas LeBien and Amanda Moon for their extraordinarily professional and ingenious insights in assembling and organizing the materials of this book.

NOTES

INTRODUCTION

2 *identified an interstellar object:* Avi Loeb, *Extraterrestrial: The First Sign of Intelligent Life Beyond Earth* (Boston: Houghton Mifflin Harcourt, 2021).

3 *Simple and plausible:* Jennifer Bergner and Darryl Seligman, "Acceleration of 1I/'Oumuamua from radiolytically produced H_2 in H_2O ice," *Nature* Vol: 615, P: 610-613; Darryl Seligman and Gregory Laughlin, "Evidence that 1I/2017 U1 ('Oumuamua) was Composed of Molecular Hydrogen Ice," *The Astrophysical Journal Letters*, 896:L8; Thiem Hoang and Abraham Loeb, "Destruction of Molecular Hydrogen Ice and Implications for 1I/2017 U1 ('Oumuamua)," *The Astrophysical Journal Letters, 899:L23*; Thiem Hoang and Abraham Loeb, "Implications of Evaporative Cooling by H_2 fir 1I/'Oumuamua," Preprint Draft Version 3/24/2023, submitted to *The Astrophysical Journal Letters*.

8 *"universe by human initiative":* Edward Farhi, Alan Guth, and Jemal Guven, "Is It Possible to Create a Universe in the Laboratory by Quantum Tunneling?" *Nuclear Physics B* 339, no. 2 (July 30, 1990): 417–90.

CHAPTER 1: ASCENDING THE LADDER OF CIVILIZATIONS

14 *At a press briefing:* Press briefing by Press Secretary Karine Jean-Pierre and National Security Council Coordinator for Strategic Communications John Kirby, February 13, 2023, the White House.

14 *"All-domain Anomaly Resolution Office":* "DoD Announces the Establishment of the All-domain Anomaly Resolution Office," US Department of Defense, July 20, 2022, https://www.defense.gov/News/Releases/Release/Article/3100053/dod-announces-the-establishment-of-the-all-domain-anomaly-resolution-office/.

18 *Voyager 1 is about:* NASA, accessed February 17, 2022, https://nssdc.gsfc.nasa.gov/nmc/spacecraft/display.action?id=1977–084A.

19 *Antikythera Mechanism:* "The Antikythera Mechanism," NASA, accessed January 9, 2023, https://science.nasa.gov/antikythera-mechanism.

21 *the twelve-inch gold-plated copper disk:* "Voyager: The Golden Record Cover," NASA, accessed February 14, 2022, https://voyager.jpl.nasa.gov/golden-record/golden-record-cover/.

22 *cost of $865 million:* "Voyager—Fact Sheet," NASA, accessed October 10, 2022, https://voyager.jpl.nasa.gov/frequently-asked-questions/fact-sheet/.

23 *UFOs, Reid:* Harry Reid, "What We Believe about U.F.O.s," *The New York Times,* May 21, 2021, https://www.nytimes.com/2021/05/21/special-series/harry-reid-ufo.html.

25 *with an extraterrestrial society:* Avi Loeb, "Mind the Gap: Cosmic Stoicism," February 16, 2022, https://avi-loeb.medium.com/mind-the-gap-cosmic-stoicism-f38dad893596.

27 *recalled derision from colleagues:* "Navy Pilots Recall 'Unsettling 2004 UAP Sighting,'" YouTube, accessed August 8, 2022, https://www.youtube.com/watch?v=ygB4EZ7ggig.

27 *"keep many observers silent":* "Preliminary Assessment: Unidentified Aerial Phenomena," National Intelligence, accessed October 26, 2022, https://www.dni.gov/files/ODNI/documents/assessments/Prelimary-Assessment-UAP-20210625.pdf.

28 *"understanding this threat":* "Preliminary Assessment: Unidentified Aerial Phenomena," National Intelligence, accessed October 26, 2022, https://www.dni.gov/files/ODNI/documents/assessments/Prelimary-Assessment-UAP-20210625.pdf.

30 *"Experience teaches it":* Richard Feynman, "What Is Science?," accessed July 20, 2022, http://www.feynman.com/science/what-is-science/.

30 *a sermon about* Extraterrestrial: Rob Dobrusin, "Searching for Other Paths," accessed July 20, 2022, https://robdobrusin.com/searching-for-other-paths/.

31 *"presuppose this reality":* Rob Dobrusin, "Searching for Other Paths," accessed July 20, 2022, https://robdobrusin.com/searching-for-other-paths/.

34 *"transmedium objects":* "DoD Announces the Establishment of the All-domain Anomaly Resolution Office," US Department of Defense, accessed October 20, 2022, https://www.defense.gov/News/Releases/Release/Article/3100053/dod-announces-the-establishment-of-the-all-domain-anomaly-resolution-office/.

37 *"we don't have the technology for":* Avi Loeb, "We Need Scientific Analysis of Satellite Data on UAP," *The Hill,* December 27, 2021, https://thehill.com/opinion/national-security/587385-we-need-scientific-analysis-of-satellite-data-on-uap/.

38 *"have been declassified":* James Crowley, "More 'Difficult to Explain' UFO Sightings to Be Declassified, Says Former Trump Intel Chief," *Newsweek,* March 20, 2021, https://www.newsweek.com/former-dni-more-ufo-sightings-declassified-unexplained-1577595.

CHAPTER 2: THE DAWN OF OUR INTERSTELLAR FUTURE

41 *"systematic scientific research":* "Public Announcement," The Galileo Project, accessed October 20, 2022, https://projects.iq.harvard.edu/galileo/public-announcement.

53 *"explained as known phenomena":* The Galileo Project, accessed July 20, 2022, https://projects.iq.harvard.edu/galileo.

58 *consumer spending in the United States:* "United States Consumer Spending," Trading Economics, accessed October 20, 2022, https://tradingeconomics.com/united-states/consumer-spending.

CHAPTER 3: NEW TELESCOPES FOR EXTRATERRESTRIALS

61 *"weatherproofed enclosure":* Matthew Szenher et al., "A Hardware and Software Platform for Aerial Object Localization," *Journal of Astronomical Instrumentation,* accessed January 9, 2023, https://www.worldscientific.com/doi/10.1142/S2251171723400020.

64 *"angles with respect to observers":* Matthew Szenher et al., "A Hardware and Software Platform for Aerial Object Localization," *Journal of Astronomical Instrumentation,* accessed January 9, 2023, https://www.worldscientific.com/doi/10.1142/S2251171723400020.

65 *similarly SkyWatch:* Mitch Randall et al., "SkyWatch: A Passive Multistatic Radar Network for the Measurement of Object Position and Velocity," *Journal of Astronomical Instrumentation,* accessed October 26, 2022, https://doi.org/10.1142/S2251171723400044.

66 *Planet Labs images:* Eric Keto and Andres Watters Wesley, *Detection of Moving Objects in Earth Observation Satellite Images*, Draft Version, October 18, 2022, in collection of author.

69 *"every part of the visible sky":* "Project Summary," Vera C. Rubin Observatory, accessed July 20, 2022, https://docushare.lsst.org/docushare/dsweb/Get/Document-13936.

69 *Rubin's 500 petabyte images:* "About Rubin Observatory," Vera C. Rubin Observatory, accessed July 20, 2022, https://www.lsst.org/about.

73 *"beneficiaries of chance":* Rae Paoletta, "This Tribute Ann Druyan Wrote for Her Husband Carl Sagan Will Make You Sob," *Inverse,* November 9, 2017, https://www.inverse.com/article/38297-carl-sagan-birthday-ann-druyan-love-letter.

CHAPTER 4: THE MESSENGER

87 *A collection of scientists:* John E. Shaw, Lt. Gen, USSF, Deputy Commander, https://lweb.cfa.harvard.edu/~loeb/DoD.pdf.

88 *supposedly built in China:* A. W. Sleeswyk and N. Sivin, "Dragons and Toads: The Chinese Seismoscope of B.C. 132," *Chinese Science* 6 (1983): 6–19.

89 *"complex regions of the world":* Suzanne Baldwin, Paul Fitzgerald, and Laura Webb, "Tectonics of the New Guinea Region," *The Annual Review of Earth and Planetary Science* 40 (2012): 495–520.

96 *study by the American Automobile Association:* Matthew DeBord, "Americans Are Dangerously Overconfident in Their Driving Skills—But They're about to Get a Harsh Reality Check," *Insider,* January 25, 2018, https://www.businessinsider.com/americans-are-overconfident-in-their-driving-skills-2018-1.

96 *"compared with their average peer":* E. Zell, J. E. Strickhouser, C. Sedikides, and M. D. Alicke, "The Better-Than-Average Effect in Comparative Self-evaluation: A Comprehensive Review and Meta-analysis," *Psychological Bulletin* 146, no. 2 (2020): 118–49, https://doi.org/10.1037/bul0000218.

97 *largest one is 2I/Borisov:* "Interstellar Comet Borisov Reveals Its Chemistry and Possible Origins," NASA, accessed April 26, 2022, https://www.nasa.gov/feature/interstellar-comet-borisov-reveals-its-chemistry-and-possible-origins.

97 *meteoroid is 'Oumuamua:* Avi Loeb, "Noah's Spaceship," *Scientific American,* November 29, 2020, https://www.scientificamerican.com/article/noahs-spaceship/.

99 *"budget in stars is necessary":* Amir Siraj and Avi Loeb, *Interstellar Meteors Are Outliers in Material Strength,* Draft Version, September 22, 2022, https://arxiv.org/abs/2209.09905.

99 *"defies planetary system origins":* "Propellants," NASA, accessed November 1, 2022, https://history.nasa.gov/conghand/propelnt.htm.

CHAPTER 5: LEAVING EARTH

106 *depend on chemical propellants:* "Propellants," NASA, accessed November 1, 2022, https://history.nasa.gov/conghand/propelnt.htm.

106 *Travel time from Earth to Mars:* "Planetary Travel Time," NASA, accessed November 1, 2022, https://www.jpl.nasa.gov/edu/pdfs/traveltime_worksheet_answers.pdf.

113 *101 Authors against Einstein:* Manfred Cuntz, "101 Authors against Einstein: A Look in the Rearview Mirror," *Skeptical Inquirer* 44, no. 6 (November/December 2020), https://skepticalinquirer.org/2020/11/100-authors-against-einstein-a-look-in-the-rearview-mirror/.

113 *"political party affiliation":* Milena Wazeck, "The Truth Phalanx," *Nature Physics* 11 (2015): 518–19.

115 *I do not stand with Karl Popper:* Karl Popper, *The Logic of Scientific Discovery* (New York: Routledge, 2002); Thomas Kuhn, *The Structure of Scientific Revolutions* (Chicago: University of Chicago Press, 1996); Michael Strevens, *The Knowledge Machine—How Irrationality Created Modern Science* (New York: Liveright, 2020).

117 *Navy pilot Lieutenant Ryan Graves:* "Lex Fridman Podcast #308," YouTube, accessed August 10, 2020, https://www.youtube.com/watch?v=qLDp-aYnR1Y.

118 *the first flew about two thousand years ago:* "History of Kites," American Kitefliers Association, accessed May 23, 2022, https://www.kite.org/about-kites/history-of-kites/.

119 *"Mr. Gorbachev, tear down this wall!":* "Reagan Speech: 'Tear Down This Wall,' 1987," Gilder Lehrman Institute of American History, accessed May 5, 2022, https://www.gilderlehrman.org/history-resources/spotlight-primary-source/reagan-speech-tear-down-wall-1987.

119 *threat from outside this world:* "Address to the 42d Session of the United Nations General Assembly in New York, New York," Ronald Reagan Presidential Library and Museum, accessed February 11, 2020, https://www.reaganlibrary.gov/archives/speech/address-42d-session-united-nations-general-assembly-new-york-new-york.

122 *"advanced characteristics and performance":* "50 U.S. Code § 3373—Establishment of Office, Organizational Structure, and Authorities to Address Unidentified Aerial Phenomena," Cornell Law School, accessed November 12, 2022, https://www.law.cornell.edu/uscode/text/50/3373.

CHAPTER 6: KNOWLEDGE AND WISDOM

134 *Our lifespan:* "Average US Life Expectancy Statistics by Gender, Ethnicity & State In 2022!" Simply Insurance, accessed July 6, 2022, https://www.simplyinsurance.com/average-us-life-expectancy-statistics/#.

136 *Lieutenant Graves reported:* "Lex Fridman Podcast #308," YouTube, accessed August 10, 2020, https://www.youtube.com/watch?v=qLDp-aYnR1Y.

CHAPTER 7: SURVIVAL OF THE OPTIMISTS

148 *the Danakil Depression in Ethiopia:* Clare Wilson, "Ethiopia's Blue Volcano Burns Deadly Sulphuric Gas," *NewScientist,* May 21, 2014, https://www.newscientist.com/article/mg22229700-100-ethiopias-blue-volcano-burns-deadly-sulphuric-gas/.

148 *Klein gave the question:* Ezra Klein, "Your Kids Are Not Doomed," *The New York Times,* June 5, 2022, https://www.nytimes.com/2022/06/05

/opinion/climate-change-should-you-have-kids.html?action=click&module=RelatedLinks&pgtype=Article.

151 *outer limits of our longevity:* Andrea Estrada, "How We Age," *The Current,* September 21, 2020, https://www.news.ucsb.edu/2020/020033/aging-process.

151 *cross-generational encouragement:* Raziel J. Davison and Michael D. Gurven, "Human Uniqueness? Life History Diversity among Small-scale Societies and Chimpanzees," *PLoS ONE* 16, no.2 (January 27, 2021) https://gurven.anth.ucsb.edu/sites/default/files/sitefiles/papers/davison_gurven2021.pdf.

157 *"happens to know physics":* Nitasha Tiku, "The Google Engineer Who Thinks the Company's AI Has Come to Life," *The Washington Post,* June 11, 2022, https://www.washingtonpost.com/technology/2022/06/11/google-ai-lamda-blake-lemoine/.

160 *"from a scientific perspective":* "NASA to Set Up Independent Study on Unidentified Anomalous Phenomena," NASA, accessed June 21, 2022, https://www.nasa.gov/feature/nasa-to-set-up-independent-study-on-unidentified-anomalous-phenomena/.

160 *"understanding of UAP forward":* "NASA to Set Up Independent Study on Unidentified Anomalous Phenomena," NASA, accessed June 21, 2022, https://www.nasa.gov/feature/nasa-to-set-up-independent-study-on-unidentified-anomalous-phenomena/.

160 *"this report will be shared publicly":* "NASA to Set Up Independent Study on Unidentified Anomalous Phenomena," NASA, accessed June 21, 2022, https://www.nasa.gov/feature/nasa-to-set-up-independent-study-on-unidentified-anomalous-phenomena/.

160 *"That's what we do":* "NASA to Set Up Independent Study on Unidentified Anomalous Phenomena," NASA, accessed June 21, 2022, https://www.nasa.gov/feature/nasa-to-set-up-independent-study-on-unidentified-anomalous-phenomena/.

161 *"with a sense of humility":* "NASA Forms Independent Team to Study Unexplained UFO Sightings," *The Guardian,* June 10, 2022, https://www.theguardian.com/science/2022/jun/09/nasa-study-ufo-sightings.

162 *atomism is traceable:* Carlo Rovelli, *Reality Is Not What It Seems* (New York: Riverhead Books, 2017): 18–20.

164 *"six million gigabytes of data per year":* "The Legacy Survey of Space and Time (LSST)," National Accelerator Laboratory, accessed July 22, 2022, https://www.lsst.org/sites/default/files/documents/rubinobs_lsst_factsheet_9_2020_final.pdf.

CHAPTER 8: OUR TECHNOLOGICAL FUTURE

167 *"another 4,106 lives saved":* "Advanced Driver Assistance Systems—Data Details—Injury Facts," NSC, accessed June 28, 2022, https://

injuryfacts.nsc.org/motor-vehicle/occupant-protection/advanced-driver-assistance-systems/data-details.

168 *four hundred crashes with cars:* Tom Krisher, "U.S. Report: Nearly 400 Crashes of Automated Vehicles," *AP News,* June 15, 2022, https://apnews.com/article/self-driving-car-crash-data-ae87cadec79966a9ba56e99b4110b8d6.

168 *a* Washington Post *op-ed:* "The Problem with Self-driving Cars? Many Don't Drive Themselves," *The Washington Post,* June 26, 2022, https://www.washingtonpost.com/opinions/2022/06/26/problem-with-self-driving-cars-many-dont-drive-themselves/.

170 *1.53 in 2020:* "Car Crash Deaths and Rates," NSC, accessed June 27, 2022, https://injuryfacts.nsc.org/motor-vehicle/historical-fatality-trends/deaths-and-rates.

170 *"not yet know which":* "'The Best or Worst Thing to Happen to Humanity'—Stephen Hawkins Launches Centre for the Future of Intelligence," University of Cambridge, accessed November 15, 2022, https://www.cam.ac.uk/research/news/the-best-or-worst-thing-to-happen-to-humanity-stephen-hawking-launches-centre-for-the-future-of.

171 *idea of the doomsday argument:* Sam Harris, *Making Sense* (New York: Ecco, 2020): 319–51.

173 *purpose of making clothes:* Katie Hunt, "When Did Humans Start Wearing Clothes? Discovery in a Moroccan Cave Sheds Some Light," *CNN,* September 16, 2021, https://edition.cnn.com/2021/09/16/africa/clothing-bone-tools-morocco-scn/index.html.

175 *"window on all time":* Thomas Wolfe, *Look Homeward, Angel* (New York: Charles Scribner's Sons, 1929): 3.

177 *period life table:* "Actuarial Life Table," Social Security, accessed June 29, 2022, https://www.ssa.gov/oact/STATS/table4c6.html.

CHAPTER 9: NOAH'S SPACECRAFT

188 *multicellular animals:* "How Did Multicellular Life Evolve?" Astrobiology at NASA, accessed July 7, 2022, https://astrobiology.nasa.gov/news/how-did-multicellular-life-evolve/.

190 *endosymbiosis, or the union of two cells:* Manasvi Lingam and Avi Loeb, *Life in the Cosmos: From Biosignatures to Technosignatures* (Cambridge, MA: Harvard University Press, 2021): 190–95.

190 *fifty most influential women in science:* Kathy Svitil, "Lynn Margulis," in "The 50 Most Important Women in Science," *Discover*, accessed January 1, 2023, https://www.discovermagazine.com/the-sciences/the-50-most-important-women-in-science.

193 *approximately 375 wars:* "20th Century Wars Timeline," The Timeline Geek, accessed July 9, 2022, https://www.thetimelinegeek.com

/wars-20th-century/; "Wars of the 20th Century," 20th Century Wars, accessed July 9, 2022, https://20thcenturywars.com.

194 *total deaths from COVID-19:* "WHO Coronavirus Dashboard," WHO, accessed December 12, 2022, https://covid19.who.int.

194 *have gone extinct:* Hannah Ritchie, Fiona Spooner, and Max Roser, "Biodiversity," OurWorldInData.org, accessed July 11, 2022, https://ourworldindata.org/biodiversity.

196 *half of all Americans are religious:* "Living Facts," Pew Trust, accessed July 12, 2022, https://www.pewtrusts.org/en/projects/archived-projects/living-facts.

196 *exists on other planets:* Courtney Kennedy and Arnold Lau, "Most Americans Believe in Intelligent Life beyond Earth, Few See UFOs as a Major National Security Threat," Pew Research Center, June 30, 2021, https://www.pewresearch.org/fact-tank/2021/06/30/most-americans-believe-in-intelligent-life-beyond-earth-few-see-ufos-as-a-major-national-security-threat/.

200 *"your sheer existence":* Rainer Maria Rilke, "Night," A Year with Rilke, November 8, 2011, http://yearwithrilke.blogspot.com/2011/11/night.html.

CHAPTER 10: THE COSMIC LADDER

202 *nonrecycled garbage:* Valentina Fernandez, "Where Does Your Recycling Really Go?" *Dartmouth Undergraduate Journal of Science,* Summer 2021, https://sites.dartmouth.edu/dujs/2021/09/09/where-does-your-recycling-really-go/.

202 *Great Pacific Garbage Patch:* "Great Pacific Garbage Patch," *National Geographic*, accessed July 12 2022, https://education.nationalgeographic.org/resource/great-pacific-garbage-patch.

203 *our Solar system formed:* "How Our Solar System Was Born," Natural History Museum, accessed, July 14, 2022, https://www.nhm.ac.uk/discover/how-our-solar-system-was-born.html.

203 *The Cambrian Explosion:* "The Cambrian Explosion," Understanding Evolution, accessed July 14, 2022, https://evolution.berkeley.edu/the-arthropod-story/meet-the-cambrian-critters/the-cambrian-explosion/.

203 *"worldwide in a single year":* "Plastic Bottles," Habits of Waste, accessed July 14, 2022, https://habitsofwaste.org/call-to-action/plastic-bottles/.

203 *"pore of productive life":* Norman Mailer, *Conversations with Norman Mailer,* ed. J. Michael Lennon (Jackson and London: University Press of Mississippi, 1988): 321.

203 *four fundamental forces:* "4 Fundamental Forces," Math Is Fun, accessed July 2022, https://www.mathsisfun.com/physics/force-types.html.

204 *Evidence of cancer:* "History of Cancer," The Cancer Atlas, accessed July 2022, https://canceratlas.cancer.org/history-cancer/.

204 *correlation with human diseases:* Stephanie L. Wright and Frank J. Kelly, "Plastic and Human Health: A Micro Issue?" Environmental Science & Technology 51, no. 12 (June 2017): 6634–47, https://pubmed.ncbi.nlm.nih.gov/28531345/.

205 *"advice are dead by now":* Avi Loeb, "Webb's Deep Insight: Time Is of the Essence," *The Hill,* July 16, 2022, https://thehill.com/opinion/technology/3561337-webbs-deep-insight-time-is-of-the-essence/.

207 *Martin Elvis, published a paper:* Martin Elvis, "Research Programmes Arising from 'Oumuamua Considered as an Alien Craft," *International Journal of Astrobiology* 21, no. 2 (April 1, 2022): 63–77, https://www.cambridge.org/core/journals/international-journal-of-astrobiology/article/abs/research-programmes-arising-from-oumuamua-considered-as-an-alien-craft/2F0CD6A0EF109B7445A11CB56035D014.

211 *"playing a virtual reality video game":* Passant Rabie, "Perseverence: Meet the Driver Who Navigates the Mars Rover," *Inverse,* May 9, 2021, https://www.inverse.com/science/vandi-verma-nasa-rover-driver.

213 *As John Chapman knew:* "Who Was Johnny Appleseed?" History.com, accessed July 2022, https://www.history.com/news/who-was-johnny-appleseed.

213 *Could machines do the same:* David Emmite, James A. Reggia, and Moshe Sipper, "Go Forth and Replicate," *Scientific American,* February 1, 2008, https://www.scientificamerican.com/article/go-forth-and-replicate-2008-02/.

213 *a video of their effort:* "Automatic Mechanical Self Replication (Part 1), Lionel Penrose, 1958," YouTube, accessed July 2022, https://www.youtube.com/watch?v=12HfFvjSAHc.

214 *"meaningful on a spiritual level":* Rob Dobrusin, "Searching for Other Paths," September 11, 2021, https://robdobrusin.com/searching-for-other-paths/.

CONCLUSION

225 *hypothesized that dark matter:* Avi Loeb and Neal Weiner, "Cores in Dwarf Galaxies from Dark Matter with a Yukawa Potential," October 15, 2018, https://arxiv.org/pdf/1011.6374.pdf.

INDEX